ROUTLEDGE LIBRARY EDITIONS: COLD WAR SECURITY STUDIES

Volume 12

CHEMICAL WEAPONS

CHEMICAL WEAPONS

Destruction and Conversion

SIPRI

Routledge
Taylor & Francis Group
LONDON AND NEW YORK

First published in 1980 by Taylor & Francis Ltd

This edition first published in 2021
by Routledge
2 Park Square, Milton Park, Abingdon, Oxon OX14 4RN

and by Routledge
605 Third Avenue, New York, NY 10017

Routledge is an imprint of the Taylor & Francis Group, an informa business

British Library Cataloguing in Publication Data
A catalogue record for this book is available from the British Library

ISBN: 978-0-367-56630-2 (Set)
ISBN: 978-1-00-312438-2 (Set) (ebk)
ISBN: 978-0-367-62698-3 (Volume 12) (hbk)
ISBN: 978-1-00-311040-8 (Volume 12) (ebk)

Publisher's Note
The publisher has gone to great lengths to ensure the quality of this reprint but points out that some imperfections in the original copies may be apparent.

Chemical Weapons: Destruction and Conversion

sipri

Stockholm International Peace Research Institute

Taylor & Francis Ltd
London
1980

First published 1980 by Taylor & Francis Ltd
10-14 Macklin St, London WC2B 5NF

Distributed in the United States of America by
Crane, Russak & Company, Inc.
3 East 44th Street, New York, N.Y. 10017
and in Scandinavia by
Almqvist & Wiksell International
26 Gamla Brogatan
S-101 20 Stockholm, Sweden

British Library Cataloguing in Publication Data

Stockholm International Peace Research Institute
Chemical weapons.
1. Chemical warfare
2. Disarmament
I. Title
358'.34 UG447
ISBN 0-85066-199-4

Printed and bound in the United Kingdom by
Taylor & Francis (Printers) Ltd, Rankine Road,
Basingstoke, Hampshire RG24 0PR

Preface

Negotiations for a convention prohibiting chemical weapons have been going on for a number of years in various forums, but no tangible progress has been achieved so far. Among the major problems to be resolved are those concerning the destruction or conversion of existing stockpiles of chemical warfare agents and munitions as well as of the production plants for them.

In order to discuss these and related subjects, SIPRI organized an international symposium in Stockholm in June 1979 which was attended by a number of eminent scholars, who are thanked for their contributions.

The conclusion arrived at by SIPRI is that adequate means and methods are available or can be developed for the elimination of all chemical weapons and relevant facilities. Consequently, the problem of destruction or conversion should not constitute an obstacle to an agreement on chemical weapons disarmament.

This book was edited by a group convened by SIPRI which included the following members: Jozef Goldblat, Arne Kjellstrand, Rajesh Kumar (also symposium rapporteur), Karlheinz Lohs (consultant), Johan Lundin (consultant), Theodor Nemec and Julian Perry Robinson (consultant). The editorial assistance of Melinda Öjermark is also acknowledged.

October 1979

Frank Barnaby
Director

Contents

PART I

Introduction

SIPRI

The question of eliminating chemical weapons through international treaty has been on the agenda of the world's principal disarmament negotiating body for the past 12 years. Concrete results are evidently some way into the future still, but a fuller understanding has been attained of the political, military, legal and other technical problems that must be surmounted in order to achieve such a treaty. These problems have engaged the attention and research resources of SIPRI since 1968. With a view both to stimulating progress and to increasing the amount of pertinent information and analysis readily available, SIPRI has published numerous technical studies of chemical disarmament matters.[1] The present volume is a continuation of these efforts. It draws from a symposium on chemical weapons convened for the purpose by SIPRI during June 1979. Most of the participants were members of the Pugwash Chemical Warfare Study Group.[2]

[1] The principal SIPRI publication in the field has been *The Problem of Chemical and Biological Warfare* (Stockholm: Almqvist & Wiksell) published in six volumes during 1971–75. Parts of this in draft form had previously been distributed in a mimeographed "provisional edition" (February 1970) to Geneva delegations and foreign ministries. The 1968/69 and 1969/70 editions of *SIPRI Yearbook of World Armaments and Disarmament* (Stockholm: Almqvist & Wiksell) had contained chapters on, respectively, chemical/biological warfare technology and CBW disarmament. The 1972, 1973, 1974 and 1975 editions of *World Armaments and Disarmament: SIPRI Yearbook* (Stockholm: Almqvist & Wiksell) contained chapters reviewing the course of the previous years' chemical talks, thereby updating volume 4 of the main study. Since then SIPRI has arranged for the publication of studies in the field commissioned from outside experts or visiting scholars, and has convened or assisted in the organization of a number of symposia, publishing volumes based on their proceedings. In chronological order, these publications are as follows: *Chemical Disarmament: Some Problems of Verification* (Stockholm: Almqvist & Wiksell, 1973); *The Effects of Developments in the Biological and Chemical Sciences on CW Disarmament Negotiations* (Stockholm: SIPRI, 1974); *Chemical Disarmament: New Weapons for Old* (Stockholm: Almqvist & Wiksell, 1975); *Delayed Toxic Effects of Chemical Warfare Agents* (Stockholm: Almqvist & Wiksell, 1975); *Medical Protection against Chemical-Warfare Agents* (Stockholm: Almqvist & Wiksell, 1976); "Dioxin: a potential chemical-warfare agent", in *World Armaments and Disarmament: SIPRI Yearbook 1977* (Stockholm: Almqvist & Wiksell, 1977), pages 86–102; "The destruction of chemical warfare agents", in *World Armaments and Disarmament: SIPRI Yearbook 1978* (London: Taylor & Francis, 1978), pages 360–76; and "Stockpiles of chemical weapons and their destruction", in *World Armaments and Disarmament: SIPRI Yearbook 1979* (London: Taylor & Francis, 1979), pages 470–89.

[2] This is a group of scientists organized via the Pugwash Council from a still growing number of countries, now 22, each of its members having a particular interest or specialization in aspects of chemical-warfare (CW) defence or arms limitation. Established in 1973, the Study Group has organized a succession of international round-table conferences and workshops on CW matters, of which the latest—the 7th Workshop—was held concurrently with the SIPRI symposium. For the final report of this workshop, and for a listing of the participants and the papers presented, see *Pugwash Newsletter*, vol. 17, no. 1–2 (July–October 1979), pages 40–48.

The subjects of the papers presented at the symposium reflected the state of the intergovernmental talks on chemical weapons as of early 1979. Not directly addressed in the present volume, therefore, are several other topics that are no less relevant to the content of a future chemical weapons convention: topics that no longer appeared to be especially controversial, or ones that had already been treated in detail in prior work. They formed a part of the background to the symposium with which most of the participants were familiar, shaping both the agenda and the discussions.

The scope of the present volume is limited also by constraints on space. In keeping with the primary focus of the symposium—problems of destruction and conversion—not all of the papers presented by participants have been included, and some have been shortened or revised so as to avoid overlaps. Nor have all the considerations introduced by participants during the symposium been reflected. The objective has been a reasonably homogeneous volume that can stand on its own as a contribution to the current international debate on the prevention of chemical warfare. The relevance to this debate of the subjects addressed perhaps needs little explanation. Nevertheless it may be useful here to describe something of their context so as to set the different papers within a common perspective.

Several reviews are available which describe the course of the intergovernmental chemical-warfare (CW) talks thus far.[3] It is enough to note here that the talks have proceeded through three distinct phases. Starting within the former Eighteen Nation Disarmament Committee (ENDC), chemical weapons were at first considered jointly with biological weapons, since the use of either is outlawed by the same international treaty—the 1925 Geneva Protocol; however, some states party have reserved the right to retaliate in kind, thereby legitimizing possession of the weapons. The first phase of the talks ended with agreement within the disarmament negotiating body—by then expanded into the Conference of the Committee on Disarmament (CCD)—to negotiate separate treaties on the two classes of weapon. Conclusion of the 1972 Biological and Toxin Weapons Convention soon followed; among its provisions is an obligation upon states party to continue negotiations "in good faith" on chemical weapons. The second phase in the talks ended in July 1974 when, unexpectedly, the communiqué from the Moscow summit meeting between President Nixon and General Secretary Brezhnev announced that the two sides would consider a "joint initiative" on chemical weapons within the CCD. This meant that for practical purposes the CCD ceased to have the primary negotiating role, this now passing to

[3] For a documented account of the beginnings of the present series of talks, and on prior talks back to 1920, see SIPRI, *CB Disarmament Negotiations, 1920–1970*, volume 4 in the series *The Problem of Chemical and Biological Warfare* (Stockholm: Almqvist & Wiksell, 1971). As noted above, the talks during 1970–74 are reviewed in the SIPRI Yearbooks for the years in question. The United Nations Secretariat has also published documented reviews covering, at the time of writing, the period up to 1977 inclusive (UN, 1970, 1976, 1977, 1978). For the period up to mid-1979, see appendix 2.

whatever private discussions the USA and the USSR might arrange between themselves; these commenced in August 1976. The third phase thus became characterized by the CCD essentially marking time on chemical weapons until disclosure of the "joint initiative", but keeping the issue alive through continued discussion of technical details and exchanges of view. This phase is still continuing. However, in the summer of 1977 an important development took place when agreement was finally reached within the US Government on the detailed content of a chemical weapons treaty judged worth pursuing by the United States (US Congress, 1979). In public this development was marked by an increase in the duration of meetings of the bilateral US–Soviet working group preparing the "joint initiative". (The paper by Goldblat, on pages 157–64, prepared after the symposium and thus presented in part III of this volume, describes what is publicly known about the status of the bilateral negotiations as of summer 1979.) Meanwhile, increasing pressure has developed within the Geneva disarmament committee—now further expanded and reconstituted as the 40-nation Committee on Disarmament (CD)—for it to reassert its multilateral negotiating role in the chemical talks.

Progress in the talks is, of course, intimately dependent upon the wider environment of international affairs. The Strategic Arms Limitation Talks (SALT), the future of détente, and inflexions in North–South relations must inevitably exert a powerful effect. Within this context, however, it is possible to analyse the particular issues which negotiations on chemical weapons must resolve. The avowed objectives of the talks would provide a convenient point of departure for such an exercise. Explicit and broadly representative statements of objective are to be found in the draft chemical weapons conventions that have so far been put forward: the Socialist countries' draft of March 1972 (CCD/361), the non-aligned countries' outline draft of April 1973 (CCD/400), the Japanese draft of April 1974 (CCD/420) and the British draft of August 1976 (CCD/512). Though the drafts do not specify identical goals, they concur in seeking a treaty which would bind all states that possess chemical weapons to disarm themselves of them and which would obligate all states to refrain from acquiring chemical weapons. The declared objective is, in short, a comprehensive disarmament/non-armament treaty on chemical weapons.

Thus of the many detailed matters requiring negotiation, a distinction may be drawn between those of disarmament and those of non-armament: those relating to *elimination* of existing CW capability and those relating to *preclusion* of future capability. There are then two principal questions which differentiate the substantive issues to be resolved. First, what is it precisely that is to be eliminated now and precluded? The broad statements of objective in the draft conventions noted above refer to "chemical agents" or "chemical weapons". But how are these to be defined to the satisfaction of everyone? High explosives are, after all, chemicals; so are flame and smoke agents; so are tear-gases; so are herbicides. These have all been used in war as weapons, but which of them are now to be proscribed as "chemical weapons"? And once this delimitation question has been resolved, there is the further question of specifying the activities associated with chemical weapons that are to be proscribed. Few states will be content with a ban solely on possession, for activities such as research, development, training, production, trade or other international transfer could appear no less threatening than actual stockpiling when conducted by potential adversaries. All of these issues—of agent/weapon and activity delimitation—may be categorized together as questions of *scope*.

Second, how are states party to assure themselves and one another that the provisions of the treaty are being complied with? The minimum requirement is for arrangements that significantly reduce the probability of mistaken suppositions of non-compliance. The quest must therefore be for a treaty regime which increases the flow of pertinent information to the necessary degree, but without doing so in such a way as to compromise legitimate military or industrial secrecy. How precisely this should be done raises intricate political and technical questions, which together may be categorized as questions of *verification*.

During the past 12 years of talks on chemical weapons, the millions of words that have been spoken or written amount to different permutations of position on the central issues of elimination, preclusion, scope and, pre-eminently, verification. Since these issues are interrelated through their many component issues, the permutations are enormous. This is disconcertingly evident from the analytical review prepared by the CD Secretariat—and reproduced in this volume as appendix 2—of positions taken on the different component issues by CCD delegations. The intricacies provide governments participating in the talks with virtually limitless opportunities for procrastination. But though these opportunities may reduce the rate of progress towards a chemical weapons convention, it should be noted that they may also be valuable during the difficult processes of reaching intragovernmental and other political accommodations at home and within alliances on the terms of an acceptable treaty.

The present volume addresses the issues of scope and verification through a focus on how elimination might actually be conducted. While this approach has the limitation of side-stepping scope and verification problems peculiar to preclusion, its emphasis on the practicalities of elimination introduces important, but hitherto absent, elements into the debate. The principal themes can be stated as follows. For the weapons and associated activities that are to be encompassed within the scope of a chemical disarmament treaty, a range of different elimination methods can be envisaged, broadly classifiable as destruction or conversion. The methods will vary in their technical, economic, public-safety, environmental-impact and other such characteristics, so that some will be more attractive than others on practical grounds. There may be variations in this respect from country to country. The methods will also differ in the extent to which they lend themselves to different forms of monitoring and control. These considerations may interact to limit the range of feasible solutions to verification questions.

The scope of the treaty should certainly be broad enough to prohibit possession or retention of those components of CW capability whose existence may be seen by potential states party as a significant threat to their security. In respect of these components the disarmament regime established by the treaty must provide both for their elimination and for adequate verification of their elimination. The paper by Perry Robinson (pages 9–56) classifies the types of chemicals, associated equipments and associated activities which together would constitute a full capability for waging chemical warfare. The paper identifies the more critical components and then collates information from open sources about the nature of some of these components—both equipments and facilities—actually available to NATO and allegedly available to the WTO. The paper thus affords a background to the shorter but more detailed presentations which follow it on the elimination of particular categories of capability component. The paper by Mikulak (pages 57–66) addresses, apparently for the first time in the open literature, the problems involved in dismantling facilities hitherto used for large-scale manufacture of CW agents and their loading into munitions. The account is based on existing US facilities for sarin nerve-gas. As to elimination of stockpiles, the paper by Lohs (pages 67–75) sets out several possibilities for converting CW agents into useful starting materials and intermediates for civilian manufactures. The paper also considers destruction methods, particularly in the light of historical experience in the disposal of stocks remaining from World War II. Such experience as regards past Japanese stocks is described in the paper by Kurata (pages 77–93). The technical, medical, public-safety and environmental-protection problems which these ageing and discarded remnants of World War II have presented—and which they continue to present in several countries, such as the Soviet Union (*Krasnaya Zvezda*, 1979) and the Federal Republic of Germany (*Frankfurter Rundschau*, 1979)—are identical with those that must be solved in any future stockpile elimination procedures. The same is, of course, also true for the operations that have been under way in the United States for the disposal of obsolete, surplus or subquality chemical agents and munitions of more recent vintage; but since these operations are already copiously detailed in the open literature,[4] they are not described further in the present volume. The toxicological dimension of the safety requirements is set out in the paper by Vojvodić and Binenfeld (pages 95–106), and the biomedical dimension in the paper by Rosival (pages 107–111). Recent

[4] It seems that about 80 per cent of the 150 000 tonnes or so of CW agents manufactured by the United States over the past four decades have been disposed of in ocean-dumping, land-burial, open-pit-burning, plant-incineration and plant-detoxification operations. Full details are available in the public record on disposal operations since the late 1960s, for this has been required by the environmental-protection legislation enacted then. SIPRI (1975, 1978) has reviewed much of this literature, to which, however, there have been several subsequent additions (e.g., US Department of the Army, 1977; Crist, 1979). From these and related sources it appears that, since 1967, the total quantity of CW agents disposed of or scheduled, as of 1977, for destruction has amounted to some 7 500 tonnes; but little detailed information is available on how the other 112 000 tonnes were disposed of during the 20-odd years before 1967.

research findings bearing upon the insidious and hitherto under-appreciated possibility of long-term sequelae of exposure to minute amounts of organophosphorus compounds, such as nerve gases, are reviewed in the paper by Bošković and Kušić (pages 113–16).

It is in the types of monitoring to which they lend themselves that the practicalities of elimination may most influence the options for verification. Verification of the implementation of treaty provisions may conveniently be thought of in terms of three contributing processes: monitoring, assessment and evaluation. *Monitoring* concerns the collection of physical data originating either in elimination activities, such as stockpile destruction, or in activities that might be mistaken for proscribed ones. The data may be of many varieties: the rate of flow of munitions into a demilitarization/detoxification plant, for example, or the intensities of particular infra-red wavelengths in light reflected from the surface of a weapons proving ground, or the amount of product acrylonitriles leaving a factory that uses hydrogen cyanide as a starting material. *Assessment* concerns the interpretation of such data: the process, first, of inferring information about the nature of the activities from which the data have emanated and, second, of judging from such inferences and any other pertinent information the probabilities of the activities under observation actually being—wholly or in part—what they appeared to be. The likelihood of any such interpretation resulting in an assessment of probability approaching unity would depend upon the types of data monitored and the capacity available for monitoring them. The highest-grade data would usually only be accessible on site, so that if on-site monitoring were not possible the probabilities thus assessed could be far short of unity, meaning a high degree of uncertainty about the activity being monitored. *Evaluation* concerns the appraisal of the implications which the assessed probabilities, taken together, may carry for future courses of action. Uncertainty, should it arise, means a possibility of abrogation: in the light of what is perceived about the broad intentions of the state in question, is it a real possibility? And if abrogation were happening to the extent that the degree of uncertainty could admit, would it seriously endanger overall security, having regard to any countermeasures already available, including the state of anti-CW protection?

Of these three component processes, that of evaluation would necessarily be conducted by each state party in respect of potentially hostile states. For this purpose each state party would have a capacity of its own, through what are commonly termed "national technical means", for performing at least some monitoring and assessment. But unless these "national technical means" amounted to an extraordinarily large and technologically sophisticated intelligence apparatus, one that could readily be focused on foreign activities of CW concern, there might easily be critical uncertainties that they could never resolve to an adequate degree. Supplementary aids to verification therefore appear to be necessary. This much has long been accepted in the intergovernmental talks, but the form which the aids should take remains deeply controversial. There is agreement that control to be provided for in the convention should include both national and international measures. The national measures would be arranged by each state party, while international measures would comprise some sort of international body functioning as a contact point between the national authorities and further charged with responsibilities of its own. The question of whether and, if so, what types of monitoring tasks should be included among these further responsibilities is especially contentious, primarily because the type of monitoring most productive for verification purposes would require the presence of foreign personnel at sensitive locations.

Solutions to this problem of intrusiveness and derogation of national sovereignty are currently being sought in focused systems of verification that minimize the role of international on-site inspection. The focusing device is to require states party to declare any stocks of chemical weapons that they may possess and the facilities that they had formerly assigned to chemical-weapons activities—production is the activity mainly being considered in this regard—and then to accept intrusive techniques for monitoring their elimination. This is the context of the three papers presented in this volume that deal with verification: the papers on stockpile elimination by Reutov and Babievsky (pages 117–21) and by Ooms (pages 123–28), and on the elimination of production plant for CW agent by Roberts (pages 129–38).

Progress towards international agreement on any topic is inevitably paced by the processes of reaching intragovernmental agreement within the countries whose delegations dominate the negotiations. This means that an analysis of the substantive issues under negotiation may not in fact afford much of a guide to future progress, for the domestic processes of political accommodation and democratic expression may turn on quite unrelated issues, ones that are peculiar to current climates of domestic politics. Moreover, the substantive issues are, in many cases, too technical to engage widespread public and élite attention, so that, in the national decision-making processes associated with the chemical talks, they may largely be subordinate to such essentially political factors as prevailing perceptions of the intentions of potential adversaries. The degree of trust and mutual confidence obtaining between parties to the international negotiations may thus become more important than the merits and demerits, objectively considered, of competing combinations of treaty provisions on elimination, preclusion, scope and verification. This is part of the logic behind the idea now being increasingly discussed of specific measures for increasing mutual trust—measures that might be taken unilaterally or multilaterally, inside or outside the provisions of the treaty. The paper by Lundin presented in this volume (pages 139–51) develops the idea and puts forward a variety of proposals on specific measures. The extended periods of time required for development, operation and completion of elimination procedures would allow ample scope for such measures to demonstrate their worth, and the treaty provisions could be phased accordingly. It is in this area that ways around the present impasse may perhaps best be sought.

References

Crist, F. H., 1979. Newly developed technology for ecological demilitarization of munitions. In: R. A. Scott, Jr (ed.), *Toxic Chemical and Explosives Facilities*. Washington, D.C.: American Chemical Society (*Symposium Series* no. 96).

Frankfurter Rundschau, 1979. [War poison discovered.] 12 July (in German).

Krasnaya Zvezda, 1979. [Tragedy did not occur.] 5 June (in Russian).

SIPRI, 1975. Methods of destruction of chemical warfare agents. In *Chemical Disarmament: New Weapons for Old*, pp. 102–13. Stockholm: Almqvist & Wiksell.

SIPRI, 1978. The destruction of chemical warfare agents. In *World Armaments and Disarmament: SIPRI Yearbook 1978*, pp. 360–76. London: Taylor & Francis.

UN, 1970. Efforts to achieve a comprehensive ban on chemical and bacteriological (biological) weapons. In *The United Nations and Disarmament 1945–1970*, pp. 349–73. New York: United Nations.

UN, 1976. Chemical and biological weapons. In *The United Nations and Disarmament 1970–1975*, pp. 141–63. New York: United Nations.

UN, 1977. Prohibition of chemical weapons. In *The United Nations Disarmament Yearbook*, volume 1: 1976, pp. 165–78. New York: United Nations.

UN, 1978. Chemical weapons. In *The United Nations Disarmament Yearbook*, volume 2: 1977, pp. 193–210. New York: United Nations.

US Congress, 1979. *Fiscal Year 1980 Arms Control Impact Statements*. Statements submitted to the Congress by the President pursuant to Section 36 of the Arms Control and Disarmament Act. Joint Committee Print (Senate Foreign Relations Committee and House Foreign Affairs Committee), pp. 218–33, on pp. 230–31. Washington, D.C.: US Government Printing Office, March 1979.

US Department of the Army, 1977. *Operation of the Chemical Agent Munitions Disposal System (CAMDS) at Tooele Army Depot, Utah*. Final Environmental Impact Statement, Office of the Project Manager for Chemical Demilitarization and Installation Restoration, March 1977.

PART II. Papers presented at the symposium

Chemical warfare capabilities of the Warsaw and North Atlantic Treaty Organizations: an overview from open sources

J. P. PERRY ROBINSON

Science Policy Research Unit, University of Sussex, UK

Abstract. Types of capability for chemical warfare (CW) are differentiated in general qualitative and quantitative terms. Data from the open literature about the principal components of offensive capability—agent/munition stockpiles, production capacity, R & D capacity and anti-chemical protection—are collated as regards WTO and NATO member states. The emphasis is on capability-in-being rather than on potential capability, although with changing technology the latter may be becoming the more important, within certain limits, thereby raising difficult CW-treaty verification problems. The review includes an account of the reliability and extent of the available open sources, having regard to the confidence-building potential, within the context of the current inter-governmental CW arms-limitation talks, of increased openness. One finding of the review is that, on the criteria of anti-chemical protective posture, production capacity and stocks of chemical weapons, NATO and the WTO appear to have comparable CW capabilities in being. Explanations of why some commentators in the West ascribe vastly superior capability to the WTO are discussed. Such perceptions of inferiority may have the effect of creating unnecessarily and unfulfillably demanding requirements for CW-treaty verification procedures. Confidence-building measures may afford a remedy.

I. Introduction

A review is offered in the following pages of what is contained in open sources on the chemical warfare (CW) capabilities of NATO and the WTO.[1]* Its primary purpose is to aid discussion of how these capabilities might best be eliminated under the terms of a future chemical weapons convention. The emphases in the review are organized accordingly. The focus is limited to the main military alliances, not because CW is wholly or even primarily their concern alone, but because the pace and directions of the current intergovernmental talks are being set by the dominant members of these alliances.

*The footnotes indicated by superscript numerals are collected together as end-notes on pages 41–44.

The secondary purpose of the review is to facilitate wider public understanding of some of the broader policy issues affecting the intergovernmental CW talks. Arms limitation is one route whereby states can seek security against the threat of chemical warfare. As such it is competitive—though in part it may also be complementary—with the military routes of defence, dissuasion or deterrence. As the talks have proceeded, this competitiveness has imposed upon many states, in an increasingly acute form, a requirement for major policy choices about the future allocation of resources to CW-related activities, and about the precise sort of CW convention it would or would not be worth achieving. In some states the important decisions have evidently been taken already. But in others they are still pending, and to the extent that public opinion in these countries may be influential, the availability of a critical comparative review of capabilities could improve public and parliamentary debate, thereby benefiting the outcome.

No guarantee can be offered, however, that the picture conveyed by the review is an accurate representation of reality. It is more a reflection of how reality is perceived and portrayed to the public than of what the reality in fact is. For the primary purpose of the review this may be a serious limitation, but for the secondary purpose it may have a value of its own, since security is essentially a matter of perception. The review is written in ignorance of the large volume of secret literature that exists on its subject matter. That being so, it is as well to begin with a brief discussion of the scope and reliability of the source material used.

Reliability of the available open information

The last time that any serving[2] Soviet official spoke or wrote openly about a Soviet chemical weapons capability was in 1938. Since then, all open information directly on the subject has been contained in Western or non-aligned publications or statements. The most that official WTO sources permit on this particular matter is inference. There have, for example, occasionally been speeches within the Geneva disarmament committee by representatives of the USSR, and working papers tabled by them, which can be interpreted to imply the existence of a Soviet chemical weapons capability. Likewise, articles sometimes appear in the Soviet press which allow a similar interpretation. The following quotation, for instance, is from a recent article by Gardov (1978) commenting on the CW negotiations:

> As L. I. Brezhnev declared, there is no type of armament that the USSR is not prepared to limit or prohibit on a reciprocal basis by agreement with other states, provided of course that no one's security is damaged, and under terms of total reciprocity, the states having the relevant armaments at their disposal. This proposition also applies fully to chemical weapons.

Again, while some member states have issued official statements disavowing possession of chemical weapons or intention to possess them, as in the case of the German Democratic Republic, others have remained silent on this matter.

Be that as it may, in the absence of direct information from official WTO sources, most—though not necessarily all—of what is said in this review about WTO chemical weapons could have had one of two ultimate origins: intelligence services, or the imagination of propagandists. Its provenance as between these two possibilities is rarely obvious and, for the outsider, becomes maximally confused in cases where émigrés or former spies purportedly reveal hitherto secret data. On WTO chemical weapons, therefore, the reliability of the present review must be regarded as low even though what is said may seem credible and may in fact be correct. Some of the available open sources are patently bogus and are therefore accorded zero reliability and disregarded; they include such items as *The Penkovsky Papers* (London, 1965). At the opposite extreme are acknowledged disclosures of bona fide intelligence assessments; but the data which they purvey have invariably passed through filters on their way to the public, emerging without any sort of qualification as to reliability or credibility. These are circumstances in which all manner of bias or deliberate distortion may remain undetectable.

On US chemical weapons the situation is different, since much of the secrecy surrounding the US capability has long been lifted. Moreover, the *modus operandi* of government in the USA precludes a high degree of secrecy so that, even as regards sensitive military programmes, government can proceed only with quite large quantities of management information entering the public domain, through, for example, Congressional hearings. The problem in forming a balanced picture of US CW capability thus becomes one of coping with a plethora rather than a dearth of reliable data. The risk of ending up with a distorted picture seems not dissimilar in the two cases (a matter that was perhaps insufficiently appreciated during SIPRI's preparation of *The Problem of Chemical and Biological Warfare*). For example, in publicly available information about US chemical weapons research and development (R & D), so much amplitude and variety has been evident that it is easy to suppose that actual production and deployment of the weapons, about which there is more secrecy, has been proceeding in a similar fashion. This is incorrect. It seems, in fact, to be a general characteristic of big-spending armed services that they are relatively inefficient in assimilating innovative, as opposed to imitative, technical developments in weapons (Robinson & Kaldor, 1979). The incomplete extent to which US armed forces, and perhaps those of the USSR as well, have accepted chemical weapons into military posture and doctrine is one of a number of illustrations of this phenomenon.

France, which seems to be the only other member state of the two alliances to manufacture its own chemical weapons, is almost as secretive about them as is the USSR. Thus, while there is no particular reason to doubt the reliability of the information openly available on French chemical weapons, it does not go very far.

As regards the anti-chemical protective side of CW capability, information is relatively freely available for both NATO and the WTO. The gaps in the information seem to relate mostly to deployment or to vulnerabilities in existing protection and R & D efforts to remedy them. The latter type of secrecy may possibly carry over into exaggerated portrayals of the efficiency of some protective equipments.

In sum, then, the open literature has several major deficiencies in its scope and contains much that cannot obviously be relied upon. If it is to serve as any sort of guide in discussions of policy, it must therefore be used critically and in full awareness of its limitations. Because of the sensitivity of much of its subject matter, and because of the readiness with which it lends itself to malign propaganda and the prosecution of ulterior motives, a case can be made for disregarding any information it contains which does not have reputable or official provenance. But such a quest for what might be thought of as objectivity would be irremediably compromised by the subjective nature of the judgements it would require on the quality of each and every one of the available sources. The present review makes no such attempt, apart from the one exception already noted: the exclusion from consideration of publications known beyond doubt to be bogus. It is of course true, to take extreme examples, that unverified newspaper reports carry less weight than considered pronouncements by government officials. But to exclude the former in favour only of the latter would be to place an unwarrantably high confidence in the veracity of government officials, and an unwarrantably low confidence in the probity of journalists. There can be no middle ground on this issue.

Nevertheless, it must be clearly recognized that the inclusive rather than exclusive approach followed in this review runs the risk of inadvertently publicizing and therefore lending further credence to reports that are in fact unfounded, and which may indeed originate in campaigns of calumny or unscrupulous lobbying in support of vested interest. Against this risk the author can offer the safeguard only of a meticulous citation of source materials. The sceptical reader can then refer to the originals and make whatever judgements about their reliability he or she feels competent to perform.

Increased openness as a confidence-building measure

To the extent that state A's weapons programmes are influenced by what it perceives of state B's, the level of available information about B's CW capability would find some reflection in the character of A's, and vice versa. With this action–reaction conception of arms races in mind, it is interesting to speculate on the manner in which the WTO responded to the US Army Chemical Corps' increased publicity about US programmes, following the 1955 Miller Report, during the late 1950s and early 1960s (SIPRI, 1973: 195–96; Chamberlain, 1977: 28–29). This was a period in which funding for the programmes was quadrupled by the Congress, after extensive and widely reported hearings; in which, as shortly became apparent to US journalists, a second nerve-gas factory was secretly authorized, built and operated; and in which, as is now evident from recently declassified information (US Department of the Army, 1977 *a*), the no-first-use CW policy enunciated by President Roosevelt in 1943 was essentially abandoned, not to be reinstated until President Nixon's CBW policy declaration of November 1969.[3] One may also speculate on the extent to which the shortage of reliable information about WTO CW capability at that time facilitated, even stimulated, expansion of the US one. It is now recognized that the data available to the US then about Soviet CW capability were sparse and of low reliability (Conway, 1972; Frank, 1972; Leonard, 1977) despite the alarming intelligence assessments of the matter that were disseminated with every appearance of confidence. Ignorance may exacerbate fear of technological surprise; and it may also allow, in so arcane a field as CW, inflated appraisals of adversary capability and intentions to pass unchallenged. Particularly against the present background of possible US nerve-gas rearmament, via binary munitions, considerations such as these lend importance to the confidence-building measures that have been proposed in Geneva regarding declarations of CW stockpiles and facilities (e.g., UK, 1976 *b*).

II. What constitutes a CW capability?

The UK possesses several dozen kilograms of nerve gas in bulk storage and provides sophisticated anti-chemical protection for its armed forces; some of its police forces are equipped with CS weapons, as are some army units. Belgium has a hundred or so artillery shells filled with sarin. Stocks of mustard gas are said to remain in Hungary from World War II. The Federal Republic of Germany is host to several thousand tonnes of chemical munitions owned and controlled by the USA. The list can be extended. In what sense can these countries be said to possess a CW capability, and is that sense one that is significant for the content of a CW convention?

The important thing, presumably, is whether or not a country's CW posture can be perceived as a security threat by another country. This implies thresholds, both qualitative and quantitative. Several such demarcations can be conceived: between protective equipments and weapons; between research quantities or agents and militarily significant quantities; between obsolete agents of munitions and modern ones; between first-use supplies of weapons and retaliatory-use supplies; between operating production facilities and moth-balled ones; between civilian equipments or facilities and military ones; and so on, in various combinations.

For present purposes, it seems sufficient to distinguish only (*a*) between *offensive CW capability* and *protection-only capability*, and (*b*) between *capability-in-being* and *potential capability*. In both cases, "capability" refers to the ability on the part of a country's armed forces to exploit its equipments so as to contribute significantly to overall combat performance. This implies that the equipments are possessed in a quantity commensurate both with their performance and with the total size of the armed forces; that supply procedures have been institutionalized for them; that employment doctrine has been developed and disseminated; and that supply and user units are appropriately trained.

In the case of offensive CW, possession of a capability defined in this way would amount to a capacity for using chemical weapons much as the country's armed forces might use any other type of weapon. Since the possibility also exists of chemical weapons being used in unconventional warfare, there is a point in differentiating between "regular" and "irregular" or "special purpose" capabilities. The latter might comprise no more than a dedicated unit of personnel trained and supplied for special CW missions. What exactly such a capability would have to consist of in order for its possible existence to be perceived as a significant threat would depend on the countries concerned: no generally applicable quantitative or qualitative thresholds can be specified. One may observe, however, that the activities that would be involved in the acquisition of such a capability are of a type that lie on the margins of those that one can realistically expect an intergovernmental disarmament agreement to constrain. One cannot seriously hope, for example, that a CW convention will preclude acts of terrorism involving toxic chemicals.

Within the context of, at least, NATO and the WTO, an offensive CW capability must comprise the following elements:

1. Stockpiles of CW munitions adapted both in type and quantity to currently deployed weapons systems.
2. Training capacity in support of chemical weapons employment.
3. Production capacity and associated logistical arrangements in support of (1), capable of replenishing stocks of munitions as and when they are used or become obsolete.
4. Research and development capacity in support of (3).
5. Issued protective equipments in types and quantities sufficient to shield combat units from the effects of their own chemical weapons.
6. Training capacity in support of anti-chemical protection.
7. Production capacity and logistics in support of (5), capable of replenishing protective-equipment supplies as and when they are used or become obsolete.
8. Research and development capacity in support of (7).

Only as regards element (1) would the matériel involved always be unique for an offensive CW capability. Each of the other seven elements could involve matériel, personnel or facilities that would be available as part either of a protection-only CW capability (some of element 4 and all of elements 5–8), or of civil manufacturing industry (some of element 3), or of non-CW military training (some of element 2). Such overlaps are familiar to everyone who has studied CW verification problems and do not need describing in more detail here. The ambiguities to which they give rise may necessitate specific confidence-building measures in any future CW disarmament regime (Robinson, 1979). What has to be appreciated most clearly about them is that their existence could greatly ease the task of acquiring an offensive CW capability rapidly and relatively inconspicuously. Hence the need to consider potential CW capability as well as capability-in-being.

For the present, the distinction between the two is substantial, whether seen in terms of lead times to acquisition of an operational offensive capability or in terms of the verifiability of a ban on such a capability. Thus, some of elements (3) and (4) must for the present involve certain types of dedicated facility that are either extremely costly to build and operate, as in the case of supertoxic-chemical production and munition-filling plant, or difficult to disguise, as with the large-area proving grounds equipped with instrumented test grids needed for development and proof of chemical munitions.

Yet, as time goes by, the requirements for such facilities may lessen. There is, first, the general factor that, as scientific inquiry proceeds, the total fund of applicable knowledge increases world-wide. No secret can be kept for long; and, in any case, many of the important discoveries that have advanced CW technology have resulted, and will continue to result, from civil-industrial and academic work rather than from military work. As the amount of requisite R & D thus diminishes, so too may lead times and the need for dedicated R & D facilities. Other, more specific, factors may also contribute. For example, standards of industrial hygiene and environmental protection are rising in some countries to the point where the emission controls and other containment and safety features required for civil chemical manufacture may permit supertoxic chemicals to be produced in pre-existing facilities. Binary-munitions technology, once it has been fully developed, promises a similar reduction of lead times and requisite investment. This is not the place to develop these points further; but they are an important caveat to the descriptions presented later in this review: some of the classical indicators, so to speak, of offensive CW capability—with which the descriptions are primarily concerned—are losing significance.

There is, however, a countervailing trend to be observed. To an extent that did not prevail in World War II, the probability of success or failure in war has become intimately linked, increasingly so, with capabilities-in-being at the outset of hostilities. Should there be a major war in Europe, therefore, the decisive battles would most probably be fought with pre-existing supplies of weapons. If that is true, the pre-eminent components of offensive capability become weapon stockpiles, protective equipment in issue and training: elements (1), (2), (5) and (6) in the list above. The importance of the other elements—research, development and production capacity—now seems to hinge only on the rate at which they could augment or upgrade stockpiles during the period of heightening international tension that would, it is supposed, precede a European war.

One thus returns to the question of scale. How large must a stockpile of chemical munitions be before it becomes threatening? Clearly, there can be no simple quantitative answer, even as between two potential adversaries, let alone a multiplicity. Numerous qualifications would be necessary: about employment doctrine and training, the positioning of the stockpiles, their quality, the level of anti-chemical protection arrayed against them, and so forth.

Historical experience perhaps offers some guidance. According to Prentiss (1937), Germany produced about 62 000 tonnes of CW agents of all types during World War I. Little remained unused, so that the average rate of consumption was about 1 300 tonnes per month. By the end of the war, total production capacity in Germany was about 3 900 tonnes per month. With regard to World War II, total actual and planned CW-agent production capacity in Germany as of late 1944 was about 11 000 tonnes per month (US Strategic Bombing Survey, 1945; Groehler, 1978: 294–95). Total nerve-gas production capacity in the United States during the late 1950s—i.e., before completion of the VX factory—was about 30 000 short tons per year (Hylton, 1972); this would have supported a maximum consumption rate of about 2 300 tonnes per month, and many thousands of tonnes of mustard gas were also on hand. Can one conclude from these figures that, for a European war, a stockpile of less than some thousands of tonnes of agent filled into munitions—i.e., some tens of thousands of tonnes of filled munitions—would have scant military significance?

In a recent study of CW policy alternatives prepared under contract for the US Defense Department (Carpenter *et al.*, 1977), it is assumed that, for the United States, a stockpile of 10 000 tons of deliverable CW agent would represent only a "limited offensive capability", whereas one of 40 000 tons would be a "major offensive capability". The distinction being sought was between a deterrent retaliatory capability and a war-fighting capability.

III. Anti-chemical protection

Much information is available in the open literature on the anti-chemical protection available to the armed forces of NATO and WTO member states, though as is usual for most types of military capability, details about certain of the performance characteristics of individual equipments, the size of stocks and their deployment are not readily available.

The same general types of protective equipment are in issue in both alliances: respirators, protective clothing, antidote auto-injectors, decontamination appliances, air-filtration equipment for collective or semi-collective protection inside fighting vehicles or installations, and so forth. All of the major WTO and NATO armies provide such equipment at the individual and unit level and army-wide training in its use. Quality and scale of issue vary between rather wide limits, however. Technical descriptions of many of the different protective equipments are to be found in the standard open reference works and in the patent literature; they are not reviewed here.

Organizationally, two different approaches are used to varying degrees in the armies of both alliances. One involves a vertical or centralized system of command and control for anti-chemical protection in which there are attachments of specialized chemical troops at successive levels of army organization, the chemical troops as a whole constituting a separate technical service or branch with its own command structure. The other involves a horizontal or decentralized system, with field commanders at different levels being responsible for assigning personnel from within their own commands to anti-chemical defence duties that may or may not be full-time; some at least of the personnel so assigned are then given special training in chemical-warfare schools over and above that which they will already have received as part of their basic training. In both the vertical and the horizontal systems, units of anti-chemical personnel are usually assigned protective tasks in respect of biological and radiological contamination as well as chemical: hence the designation "ABC" or "NBC" (atomic/nuclear, biological and chemical) commonly used throughout NATO.

The NBC defence organizations of NATO and WTO armies differ in the relative emphases placed upon the vertical and the horizontal approaches, which in most armies exist side by side. Some rely more on specialist NBC troops for preserving overall military effectiveness in a contaminated environment; others rely more on forces-wide capability. Which course a country follows should probably be seen as a reflection of its traditional approach to force structuring in general rather than as a measure of its attention to chemical warfare. It is interesting, but probably fruitless, to speculate on which of the two approaches would be better suited to the support of an offensive CW capability.

WTO armed forces

Dominating the WTO anti-chemical protective stance for ground warfare are the Soviet Chemical Troops (BKhV), which constitute a separate arm of the Soviet Ground Forces. In the past they have had responsibilities for special-weapons employment (Manets, 1968)—incendiary and smoke—but, as far as toxic weapons are concerned, their function nowadays is believed to be purely defensive (Finan, 1974; US Department of the Army, 1975), although some Western writers have suggested otherwise (e.g., Association of the US Army, 1976). Chemical troops are organic to all Soviet combat units, their attachments ranging from one brigade at army-group or front level to one platoon at regiment level (US Department of Defense, 1969), with smaller teams down to company level (G. S. Brown, 1978). On full establishment this would mean a total strength approaching 130 000 Chemical Troops; figures for current BKhV manning that have been quoted in Western sources are around 80 000 (Lennon, 1978).

Babushkin (1978) has provided an authoritative account of the Soviet Army Chemical Service up to the end of World War II. The primary functions of the BKhV nowadays are reconnaissance for chemical, biological or radiological contamination, and decontamination, and are frequently discussed in detail in open Soviet military journals and other publications (e.g., Velenets *et al.*, 1968). The methods used for these tasks are markedly labour-intensive when compared with, for example, West German, British or US methods, although certain new items of detection and decontamination equipment that have been entering service in recent years appear exceptional in this respect. One much-remarked example is the TMS-65 mobile decontamination system; some commentators have suggested that, in view of its apparent capacity for decontaminating tanks and other front-line fighting vehicles very rapidly, it has been designed and deployed for use as part of offensive CW capability (Hoeber and Douglass, 1978). The TMS-65 consists of a turbojet aero-engine swivel-mounted on a truck chassis; used in pairs, it is said to be capable of decontaminating vehicles as they drive past by directing a decontaminating jet exhaust at them (US Department of the Army, 1975); a tank may be decontaminated in $\frac{1}{2}$–3 minutes, so it is reported (Volz, 1976), though presumably only in the case of the more volatile contaminants. Czechoslovak ground forces have similar devices (*Soldat und Technik*, 1976 *a*; Battista, 1977). While high-capacity mobile decontamination systems are to be found in NATO service, high-speed jet-exhaust devices as yet exist only in prototype (e.g., *Army R, D & A Magazine*, 1979 *b*).

In the armed services of most of the other WTO countries, the organization, equipment and training for anti-chemical protection generally follows—or has been starting to follow—the Soviet pattern, according to Western reports (e.g., *Soldat und Technik*, 1976 *b*) and as is evident from the numerous articles on specific aspects of anti-chemical protection that are published in open WTO military journals. There appears to be more co-operation in R & D and production, and hence a greater degree of standardization, than exists in NATO (G. S. Brown, 1978). Among the WTO allies, the GDR in particular is prominent in its attention to NBC protection. A Swiss military journal has recently reported that the *Nationale Volksarmee* is expanding its Chemical Service, a chemical platoon to become organic to some regiments, with a chemical battalion at army level (*Allgemeine Schweizerische Militärzeitschrift*, 1978; *Military Review*, 1979). There are numerous open GDR military manuals and other specialist textbooks dealing with chemical warfare that are frequently updated and that reveal sophisticated attention to the scientific and technical problems of CW defence (e.g., Dehn *et al.*, 1967; Franke, 1976; Franke *et al.*, 1976; Lohs, 1974; Lohs and Martinetz, 1978; Stöhr *et al.*, 1977). Other WTO countries have also published numerous textbooks on CW defence in the open literature, including the

USSR (e.g., Aleksandrov, 1969; Drugov, 1959; Karakchiyev, 1968; Levin *et al.*, 1960; Manets *et al.*, 1971; Stepanov and Popov, 1962; Stepanskiy *et al.*, 1966; Sterlin *et al.*, 1971; Stroykov, 1970; Zhuk and Stroykov, 1972); but those of the GDR have few counterparts either in the East or in the West. A recent publication from Czechoslovakia on the scientific basis for medical protection against CW weapons (Matoušek *et al.*, 1979) should, however, be noted here.

With regard to anti-chemical equipment for the individual soldier, Soviet respirators and protective clothing are known to be efficient in shielding their wearers from toxic chemicals but to be considerably more burdensome than the best of the NATO items; comparative troop trials in the USA and elsewhere have established this (Phillips, 1976; Lennon, 1978), though the details have not been released. The current respirator is less easy to don quickly than, for example, the US M17A1 or the British S6, and is heavier and more uncomfortable. The protective clothing is made from non-breathing rubberized fabric, in contrast to the air-permeable items of the larger NATO armies: below 15°C the clothing can be worn only for about four hours before heat stress builds up to casualty levels, and above 21°C the tolerance time is less than half an hour (Lennon, 1978). Factors such as these must be regarded as limitations on overall Soviet CW capability, though it is expected that improved equipments will soon enter service (Donnelly, 1977; Hoeber and Douglass, 1978). There may be some compensation, though not without peculiar costs of its own, in the increasing deployment throughout WTO ground-forces of armoured personnel-carriers and infantry fighting vehicles that are equipped for full collective NBC protection (Brown, 1977), with seals and positive-pressure filtered air systems. There may also be compensation in the frequency of training manoeuvres and other exercises involving full anti-chemical precautions. As is now increasingly the case in NATO, much attention is paid during training to realistic simulation of CW environments (Volz, 1976), while NATO forces rely on irritants or physiologically inactive simulants for this (*Army R, D & A Magazine*, 1978). Reports exist (e.g., Binder, 1978) of Soviet battle exercises at regiment level in which nitrogen mustard at vesicating dosages has been used as training agent.

One other Soviet individual-protective item should be noted here: the auto-injectors for nerve-gas antidote. The preparation used in this—known in the West as "BAT" or "TAB" and in the WTO as, so it is said (*Army Reserve Magazine*, 1976; US Department of the Army, 1977 *b*), "Nemikol-5"—differs markedly from most NATO ones, being a formulation not only of atropine but also of benactyzine and trimedoxime. This offers a remedy against poisoning both by the standard Western nerve gases (sarin and VX) and—though to a lesser extent—by nerve gases refractory to standard Western treatments, such as soman (US Army Edgewood Arsenal, 1975). Its deployment first became widely known in the aftermath of the 1973 Yom Kippur Arab–Israeli War, when Egyptian units were found to have supplies, and has been taken as evidence of soman being a standard Soviet CW agent. The fact that soman and trimedoxime can react under physiological conditions to produce an acetylcholinesterase inhibitor even more potent than soman has led to some speculation that the Soviet agent is not racemic soman but a preparation from which the pair of diastereoisomers that react most rapidly with trimedoxime is absent. It may also be noted, however, that trimedoxime is one of the better antidotes available against tabun.

NATO armed forces

Specialist NBC troops analogous to the Soviet BKhV exist in NATO armies; the *ABC-Abwehrtruppe* of the *Bundeswehr* (on which see SIPRI, 1973: 225–29; R. Meyer, 1976; Mayer, 1979) are a prominent example. The US Army, having virtually disbanded its Chemical Corps after the 1962 reorganization in favour of what was expected to be a more closely integrated CW capability (Stubbs, 1963), has since been in the process of building it up again (Tice, 1976). As of 1971, a fully manned US field army of 390 000 men would have included 3 800 chemical troops; a further 2 600 would have been included among the 154 000 people manning its communications zone (US Department of the Army, 1971 *a*). As of 1975, there were less than 2 000 chemical officers and enlisted men having chemical military occupational specialties in the US Army (Cooksey, 1975 *a*; Tice, 1976). The number of US chemical troops has since been rising, especially with the recent re-establishment of divisional chemical companies (Association of the US Army, 1976). By the end of 1980, the total number of NBC Defense Companies to have been activated and assigned to US Army combat divisions and Corps support commands will have risen to 11, together with a total of 43 NBC Reconnaissance and Decontamination Teams; additional activations are planned further ahead (Rogers, 1979 *b*). However, even when the expansion is completed, the number of US chemical troops will remain inferior to that of the Soviet BKhV by around one order of magnitude (Hoeber and Douglass, 1978). The US Defense Department believes, nonetheless, that this disparity is more than compensated for by qualitative superiorities in equipment and technique, and has stated that it has no wish to emulate what it sees as the Soviet Union's "giant janitorial force" of BKhV (*Arms Control Today*, 1979).

Side-by-side comparisons of US and Soviet chemical troops are in any case likely to be misleading as a guide to the overall anti-chemical protective effectiveness of the two armies, since they take into account only the vertical chemical defence organization; they exclude altogether from consideration the horizontally organized protective posture. In the US Army this is in theory very substantial, though in practice its effectiveness will depend heavily upon the diligence with which individual field commanders have applied, through training and practice, current Army doctrine on NBC defence (e.g., Bay, 1978; Gibbs, 1978). Current doctrine (US Department of the Army, 1977 *c*) requires, for example, that each company commander organize an NBC Defense Team from his own resources, assigning NBC defence duties full-time to some or all of the team members if he judges it necessary. Depending on the number of NBC-contamination detection kits in issue to the company, the size of its NBC Defense Team may range from 16 to 34 men out of a total company strength of 100–130 men, 3 of them being sent for additional CW-school training. At least on paper, therefore, it would appear that, while there are no more than a few thousand specialist NBC troops in the US Army, somewhere between 90 000 and 250 000 Army personnel have at least part-time responsibilities for unit NBC defence tasks.

Since 1974, in the wake of re-evaluations made by each of the US military departments of the adequacy of its anti-chemical protective posture, an accelerated programme of US chemical-warfare preparedness has been getting under way.[4] The Army is now part-way through a five-year $1 500 million programme of additional protective-equipment procurement and training (Keith, 1978; Watson, 1978; *National Defense*, 1979). The Air Force has a $234 million programme for improving the anti-chemical defences of its European air bases due for completion in 1984 (Gilbert, 1978; see also Lenorovitz, 1979), and is in the process of establishing an extra 693 chemical-defence positions (Currie, 1978). These measures will reverse the gap that had developed during the 1960s and early 1970s between the adequacy of the US anti-chemical protective posture and those of the other leading NATO services (Haig, 1978; Meyer, 1978 *a*). This, no doubt, is part of the reason for the confident Defense Department statement quoted above concerning disparities with Soviet anti-chemical protection. Detailed side-by-side qualitative comparisons of NATO and Soviet equipments have also contributed. The latest Western protective clothing, for example, can be worn indefinitely at temperatures below about 28°C (Watson, 1978);[5] the adsorptive-carbon lining which allows their air-permeability is reported to be unlikely to become saturated in less than a period of days of continuous exposure to anticipated field concentrations of toxic agent, even in summer (Lennon, 1978). The latest US protective overgarment ensemble is designed to provide a minimum of six hours protection after being worn for 14 days (Kjellstrom, 1976: 128). It is instructive to note that part of the current US re-equipment programme includes procurement and issue of TAB antidote, though this is seen as an interim measure pending standardization of one of the improved preparations now in development and is to be supplemented by issue of prophylactic pyridostigmine (Davis, 1978 *a*, 1978 *b*). The improved antidote is said to promise twice the efficacy of TAB in countering soman poisoning, and 3.5 times the efficacy when used in conjunction with the prophylaxis (Perry, 1978: 5881–82).

Dodd (1976, 1977), *International Defense Review* (1975, 1978 *a*), Lennon (1978), *Marine Corps Gazette* (1979), Marriott (1977) and *Wehrtechnik* (1977) are some of the sources which provide descriptions of the newer protective equipments in NATO service. There are further items whose development is nearing completion which are said to offer major improvements in protective capability. One is the US XM29 multipurpose respirator (US Army Armament R & D Command, 1977; Davis, 1978 *a*); others include large-scale decontamination equipment (US Arms Control and Disarmament Agency, 1979).

NATO as a whole is currently embarked upon an alliance-wide programme to smooth out disparities in the anti-chemical protective postures of its component armed forces and to upgrade overall CW defences. This dates from, at least, early 1976, when a need for improved anti-chemical defences was among the main points emphasized in the agreed 1977–82 force goals (Hornsby, 1976). Part of the increased collaboration within NATO on CW is in R & D, part in joint purchasing of protective equipments, and part in cooperative and joint training. For the last of these, the planning organization is a permanent NBC Working Group established in February 1978 within the Euro NATO Training Group, in which all members of the alliance participate except France, Iceland and Luxembourg (H. Brown, 1978 *b*). NATO spokesmen continue to express concern in public about CW preparedness, following an initiative taken in this respect by Secretary-General Luns in June 1975 (Corddry, 1975) followed up on many subsequent occasions by the Supreme Allied Commander, General Haig (e.g., Haig, 1976; *Daily Telegraph*, 1977; Weinraub, 1977; Jacchia, 1978) and, after the May 1979 meeting of NATO Defence Ministers, by the Chairman of the NATO Military Committee, General Gundersen (e.g., Haworth, 1979; *Frankfurter Allgemeine Zeitung*, 1979). General Haig's recent successor as SACEUR, General Rogers, has been continuing the process, advocating, moreover, an expanded NATO chemical-weapons retaliatory capability (Farr, 1979 *a*).

IV. Research and development capacity

At the present state of development of CW technology, substantial resources in the way of qualified personnel and experimental facilities are required if a country's CW R & D capacity is to contribute significantly to overall CW capability. While this suggests that relatively few R & D establishments exist whose elimination or conversion to other activities needs to be considered in the context of a CW convention, the picture is complicated by the intricate networks of collaboration and sharing-out of R & D tasks among different establishments which characterize the R & D capacity of both alliances. With such a division of labour, even very small laboratories may be important. There is thus a point in identifying all such establishments, possibly even in the form of a register instituted by international convention. So substantial a task is not attempted in the present review, though the beginnings of a foundation are laid in a past publication by SIPRI (1973).

The concern here is with R & D capacity in support of offensive rather than protection-only capability. Some of the problems noted earlier thus need to be considered further. For example, most of the research and some of the development work needed to build up a chemical weapons capability are also directly relevant to protection. And, by and large, a greater range of difficult technical problems confront acquisition of efficient anti-chemical protection than they do acquisition of efficient chemical weapons. Taken together, these considerations mean that if CW weapons R & D were to be conducted, much of it could be undertaken within R & D facilities ostensibly dedicated to anti-chemical protection.[6] It is extremely difficult, therefore, to characterize any particular CW R & D establishment as being relevant only to protection capability. One may hear or read official statements about such places that disavow offensive intent, but for as long as any of the work done in them remains secret, disbelief cannot be suspended. In the context of a CW convention such problems cannot be resolved by any negotiable form of verification; the remedies lie rather in confidence building (Robinson, 1975 *b*, 1979).

That being so, it seems useful to concentrate here only on R & D establishments which possess facilities unambiguously associated with chemical weapons programmes, past or present. The most obvious examples of such facilities are:

(*a*) Pilot plant for development of large-scale CW agent production processes.

(*b*) Pilot plant for development of large-scale filling and other assembly processes for chemical munitions.

(*c*) Proving grounds having the instrumented test ranges and firing grids necessary for development and evaluation of chemical munitions.

It is at least arguably true that a comprehensive programme of anti-chemical protection R & D could need access to facilities comparable with (*a*) and (*c*) above, and conceivably also (*b*). Such argument would rest on the importance of conducting detailed threat assessments and, in the case of the proving grounds, of evaluating the performance of protective items under real challenge conditions. Scale is probably the decisive characteristic. For example, alarms could be adequately test-challenged by an agent aerosol/vapour of substantially lower concentration than actual munitions would yield near to their point of burst. But if munitions were to be tested for developmental purposes, a very much greater area of unpopulated terrain would have to be set aside: the toxic cloud from a single large nerve-gas warhead or bomb may under some weather conditions travel several tens of kilometres before becoming diluted to harmless levels. This suggests that only CW testing grounds of an area exceeding several hundred square kilometres may definitely be regarded, nowadays, as chemical weapons facilities.

WTO countries

The open literature of the WTO countries is devoid of any information bearing upon post-World War II pilot plants or test facilities of the types just described. Apart from some recently declassified US intelligence assessments from the late 1940s and early 1950s, open Western sources are either uninformative on such matters or very low in reliability.

As to pilot plants, they offer almost nothing worth recording. One possible exception, but of highly questionable reliability, relates to the GDR and comprises information published last year in FR Germany (Roberts and Freeman, 1978) alleging the development and synthesis (on unspecified scale) of certain novel types of V-agent nerve gas at VEB Arzneimittelwerk Dresden and in the Magdeburg chemical and pharmaceutical works of VEB Fahlberg-List.[7] This has been reiterated uncritically by subsequent Western commentators (e.g., Tolmein, 1978) who have not, however, offered any sort of corroboration. It may be noted that both of these factories are located within densely populated areas; for this and other reasons the allegations seem implausible.

As to proving grounds, a British working paper for the CCD refers to a Soviet test area at Shikhany (UK, 1976 *a*). This site had been remarked some years earlier in open Congressional testimony from the US Department of Defense, during which it was said to provide the location, also, of the Higher Officers Chemical School (US Department of Defense, 1969). The Shikhany facilities, near Volsk on the Volga, would seem to originate in the collaborative Soviet–German CW programme of 1928–31 which has been described in some detail by SIPRI (1971 *a*: 279–80 and 284–86). Pozdnyakov (1956), a Soviet émigré formerly a chemical officer of the Red Army, states that before World War II the Central Army Chemical Polygon (TsVKhP) was located at Shikhany. This information is also contained, though with little additional detail relevant here, in a 1949 US intelligence estimate of Soviet CBW capabilities (US Joint Chiefs of Staff, 1949).[8] This estimate records the Shikhany site as having been the location, in 1942, of a storage depot—"Military Depot no. 303"—for CW agents and protective equipments, and of a CW school at which 5–7 battalions of troops were garrisoned. Other CW test areas mentioned in the estimate include a relatively small (25-square-kilometre) site in the Zaporozhskaya oblast of the Ukraine that had been active at the end of 1945; and it refers to "a recent Swedish report" of two "motorized experimental units—so-called chemical divisions" at Alma Ata and Tashkent that were field-testing dichloroformoxime as a candidate CW agent.

The available open data on WTO offensive CW R & D are thus extremely sparse. All that can be added is the information given by the US Army to a Congressional committee four years ago that Soviet field-testing of thickened-soman munitions appeared to have been taking place on newly extended test-grids; their location was not disclosed (Augustine, 1975).

NATO countries

Some degree of coordination of CW R & D within NATO is achieved through the NBC Defence Panels of the NATO Armaments Group and Defence

Research Group. In addition, there are numerous bilateral and multilateral arrangements of various kinds between member countries (SIPRI, 1973: 202–203). Prominent among them are the quadripartite programmes involving the USA, Canada, Australia and the UK. These originate in the very close tripartite arrangements developed across the North Atlantic during World War II. The co-operation continued after the war; speaking of the USA and the UK, the author of the recent official history of the British nuclear programme states that "in chemical and biological warfare the programmes of the two countries remained so closely in step as to be virtually integrated" (Gowing, 1974). The greater part of NATO's present offensive CW capability, as well as much of its protective capability, stems from this international effort. As one prominent figure in the US CW programme has described it, the weapons concepts embodied in the capability largely originated in the UK, where much of the research and early development of the current CW agents was also done; the United States took on full-scale development and manufacture, field-testing and evaluation being divided between the large-scale facilities available in Canada and the United States. Since the 1950s, British and Canadian contributions to offensive R & D have been fading away. The key R & D establishments to have been involved are these:[9]

UK Ministry of Defence Chemical Defence Establishment, Porton Down: the UK's leading, and now sole, CW R & D establishment. CDE Porton, located in Wiltshire, was founded in 1916. After the recent heavy cut-backs (Fairhall, 1977), it has about 750 people on its staff, including some 70 degree-holding scientists, among which are biological-defence specialists transferred upon closure of MRE Porton. A few square kilometres of open-air test facilities are available.

UK Ministry of Defence Chemical Defence Establishment, Nancekuke: the process-development out-station of CDE Porton, until the decision was taken in 1976 to close it down. Located in a remote part of Cornwall since 1951, it has been the site of an automated pilot plant for sarin having a nominal capacity of 70 tonnes per year; in all, about 20 tonnes of sarin were produced, during 1953–55. As was clearly apparent during the March 1979 international site visit organized through the Committee on Disarmament (UK, 1979), CDE Nancekuke is now dead as a CW R & D establishment.

US Army Chemical Systems Laboratory, Edgewood Arsenal, Maryland: the leading US CW R & D establishment, located in Maryland and dating from 1918. It appears to have around 700 or 800 scientists and technicians on its staff. It includes agent-production and munition-filling pilot plant, and has two small open-air test ranges totalling about 5 km^2.

US Army Dugway Proving Ground, Utah: the principal US chemical weapons test site, founded in 1942 in order to provide more extensive range facilities than were available at Edgewood. Since 1969 all open-air trials have, by law, had to be conducted with simulant agents. The staff is about the same size as that of Edgewood (US Department of the Army, 1976 *b*). The total area is about 3 300 km^2.

Canada, Defence Research Establishment, Suffield: established in 1941 as a joint Anglo–Canadian CW proving ground, DRES currently has some 4 000 km^2 of terrain available for CW and other trials and training exercises. Testing of actual chemical weapons there ceased before 1959. Much of current Canadian protective R & D is done at DRES.

Few particulars of the association of non-English-speaking NATO countries with these establishments, or of their own offensive CW R & D are to be found in the open literature. FR Germany and—to an uncertain extent—Italy are precluded by international treaties from possessing offensive CW capability (on which see SIPRI, 1971 *b*: 190–219). France, as the only member of the group to have its own chemical weapons in significant quantities, is pre-eminent, and there has been some degree of collaboration between French and US chemical-weapons research, development and testing teams, which still continues.

The principal French CW research station is the *Centre d'Etudes du Bouchet*, near Paris. Up to the time of World War II, semi-industrial scale pilot-plant for CW-agent process development existed at Le Bouchet, with related facilities at Vincennes and Aubervilliers, including developmental and full-scale munition-filling plant; the principal chemical weapons proving ground then was the *Polygone d'Entressen* in the Bouches-du-Rhône, near Arles, though an enormous chemical test area had recently been opened at Beni Ounif in the Algerian Sahara (SIPRI, 1971 *a*: 290–91). The latter facility, known as *B2-Namous* and occupying some 5 000 km^2, continued to be operated until 1967; work done there included evaluation of chemical protective equipments by several other NATO countries. Open sources do not indicate where large-scale testing is done now that Beni Ounif is under full Algerian control. Nor do they disclose much about the current state of the original pilot plants; some sort of nerve-gas production facility apparently exists near Toulouse (SIPRI, 1973: 215–18; Le Hénaff, 1979).

It should be recorded here that the principal German chemical weapons R & D establishment of the Hitler (and preceding) period—the *Heeresversuchstelle* on Lüneberg Heath, comprising pilot-plant facilities at Munsterlager and 120 km^2 of test area nearby at Raubkammer (SIPRI, 1971 *a*: 280)—has now become, after a period of British occupation (and some use), the NBC Defense Research and Development Institute of the Federal Armed Forces, concerned solely with protection (Metzner, 1978; Morton, 1978). The chemical-warfare materials discovered in September 1979 at a derelict chemical factory in a suburb of Hamburg (*Der Spiegel*, 1979) have been moved to this establishment pending final disposition.

V. Production capacity

A government seeking to acquire militarily significant supplies of CW agent has two broad options: either it can purchase the chemicals on the civilian market or it can build and operate its own production plant. Changing technology both in the CW field and in normal peace-time chemical manufacture has placed changing restrictions on these options. In the days when phosgene, hydrogen cyanide and mustard gas were the dominant CW agents, civilian production capacity could in many countries be relied upon to meet requirements. With the advent of the nerve gases, the heavy investment needed in enhanced safety measures, and the poor returns that could be expected on that investment as regards the civilian market, greatly reduced the feasibility of this option, compelling those few governments that wished to do so to construct dedicated facilities (Watkins, 1968). To repeat what has been said previously, this may become less true in the future as new civilian chemical plant construction moves further towards zero-emission standards; and binary-munitions technology is also reopening the civilian option.

It is not possible here to review those features of NATO- and WTO-country civilian production capacity that must, for the foregoing reason, be considered relevant to overall CW capability. Nor will account be taken, as it should, of civilian production capacity for dual-purpose CW agents. The descriptions are limited to what is known of large-scale production plant dedicated to single-purpose CW agents.

WTO countries

The open literature of the WTO countries contains no information about dedicated CW-agent factories. The open Western literature treats WTO production capacity only in very general terms, and what is said varies widely both in substance and in reliability.

According to one of the more authoritative of recent sources, a West German one, "the Soviet Union... has a chemical industry capable of producing 30 000 tons of chemical munitions per year" (Rühle, 1977, 1978). This, as can be seen by comparison with the historical data on production capacity given on page 16 above, is a modest figure. It contrasts very strongly with, for example, the figure of 8 000 tons per month of CW agent—sufficient to yield perhaps one million tons of chemical munitions per year—which Germany estimated for Soviet production during World War II (Brown, 1968).

No indications are available as to how closely the West German source reflects current US assessments of Soviet production capacity. Such assessments are used as one of a number of bases for estimating total Soviet stockpile. A US Army witness expressed this as follows when providing a Congressional committee with such a stockpile tonnage estimate in 1975: "We estimate by what we see that they have in their production capability, but there could be more produced at maximum effort" (Cooksey, 1975 *b*).

As to the locations of Soviet CW-agent plant, US intelligence assessments of the late 1940s and early 1950s—based on what was clearly very poor quality input—refer to a large-scale tabun plant in operation "somewhere east of the Urals" (e.g., US Joint Chiefs of Staff, 1949). The current impression appears to be that the USSR manufactures CW agents within closed central areas of large civil-chemical complexes. This seems to be borne out by a recent émigré publication which states that the agents are manufactured within the civilian chemical industry (Agursky, 1976) rather than within the military sector of the Soviet industrial economy.

Other WTO countries have been reported to possess some sort of CW-agent production capacity (Black, 1961), though according to a "US–USSR CW Capability Comparison" recently released by the US Joint Chiefs of Staff, they all "rely on Soviet supplied munitions" (G. S. Brown, 1978).

NATO countries

Only France and the United States have large-scale production capacity for single-purpose CW agents. On the French capacity, all that is known from open sources is a 1970 report that the Service des Poudres factory at Pont-de-Claix specialized in "armements chimiques" and that its maximum work force numbered 1 700 people (Pergent, 1970).

The US nerve-gas production facilities, all of them built out of Defense appropriations, are currently in "layaway" status. The following details are mostly from Hylton (1972):

Muscle Shoals Phosphate Development Works, Alabama. This was designed, built and operated during the period 1950–57 for production of methylphosphonyl dichloride ("dichlor")—an intermediate for sarin. At capacity, MSPDW could have produced about 30 000 tonnes per year, though total actual production, from commencement in 1953, seems to have been in the range 11 000 to 21 000 tonnes (SIPRI, 1975). Dichlor production was, at that time, beyond the capacity of the civilian US chemical industry, but it seems from what is known of current planning for binary-munition procurement that commercial manufacturers would nowadays be willing to provide it (Robinson, 1975 *a*: 11). Much of the MSPDW plant has by now probably deteriorated to uselessness.

Rocky Mountain Arsenal, Colorado. Production facilities for sarin were built and operated at RMA during 1950–57, using dichlor as the starting material.[10] Capacity, envisaged as 9 000 tons per year in 1952 but subsequently trebled, was keyed to that of MSPDW, and it seems that from start-up in 1953 to layaway in 1957 something of the order of 15 000 to 20 000 tonnes of sarin were made in all.

Newport Chemical Plant, Indiana. Construction of production facilities for VX was completed in 1961, two years after design had commenced, on the site of an old heavy-water plant. Details on NCP capacity are not available, but the estimated 4 000–5 000 tonnes (SIPRI, 1975) of VX that were made during 1961–67 resulted from very considerably less than full-capacity operations. The four-step transesterification process from phosphorus trichloride via ethyl 2-diisopropylaminoethyl methylphosphonite ("QL") was used, exploiting *inter alia* early process studies done in the UK at CDE Nancekuke. NCP was placed in layaway status in 1969.

Should the USA decide to resume nerve-gas procurement, these three facilities might be decommissioned if the binary-munition route were to be followed. Since the US Army's original expectation of being able to purchase all the requisite binary reactants from the US chemical industry now seems unlikely to be fulfilled, it would then be necessary to construct new facilities. For binary sarin the unpurchasable—for the present—reactant is methylphosphonyl difluoride ("DF"); and in 1973 the US Army announced that it was going to build the necessary production plant itself, Pine Bluff Arsenal in Arkansas having been chosen as the site. PBA is a chemical-weapons production and storage complex dating from World War II, when plant for Levinstein mustard, lewisite and HN-1 nitrogen mustard, as well as for smoke and incendiary munitions, was set up there. Subsequent additions included, in the early 1960s, a filling line for BZ munitions, the BZ itself being purchased from industry. As for the projected DF plant there, Congress deleted the necessary initial funding from both the 1975 and the 1976 budgets; since then the request has not been resubmitted.

In addition to DF process plant, facilities for assembling binary munitions would also be built at Pine Bluff. What is planned is a "modular type facility capable of manufacturing at one site a variety of items, essentially ground-delivered and air-delivered systems, with common utilities, security and safety features" (Davis, 1978 *b*). The design is complete—largely on the basis of pilot work at Edgewood Arsenal—as regards the M687 155-mm binary sarin projectile, but it is currently being expanded (Watson, 1978) to provide for the assembly of two binary VX munitions, one being the XM736 8-inch (203-mm) projectile, the other the BLU-80/B (Bigeye) 500-lb aircraft bomb.[11] No information has yet been released on sources for binary VX reactants, namely QL and the dimethyl disulphide formulation designated "NM". A commercial source for the latter can probably be found; and if one cannot be found for the QL, the existing production line for it at Newport Chemical Plant could perhaps be refurbished.

The principal alternative to the binary route for US nerve-gas rearmament is the filling of existing supplies of nerve gas—from bulk stocks or current munitions (perhaps using the transportable Drill and Transfer System that the Army currently has under development (*Army R, D & A Magazine*, 1979 *a*)), or both—into new munitions. This would necessitate redistillation of the nerve gas (Miller and Cooksey, 1977), for which purpose plant at Rocky Mountain Arsenal or Newport Chemical Plant might perhaps be retained, though there would presumably be a case for establishing new plant for this purpose, possibly at one of the main chemical storage depots, such as Tooele Army Depot or Pine Bluff Arsenal.

VI. Agent/munition stockpiles

Open information on stockpiles is very incomplete. A certain amount is available on total levels of stocks but, for the purposes of the present review, it leaves unanswered a whole range of basic questions. How accessible are the stocks to forward commanders? Are the military capabilities which they provide evenly spread across the spectrum of deployed weapons delivery systems or would their use be significantly limited by mismatches in this respect? Are the munitions of designs that are suited to the latest delivery systems or can they be used only with obsolescent ones? Has the quality of the munitions been affected by deterioration in storage? For stocks in the form of bulk agent, how rapidly could they be filled into munitions, and are the filling lines necessary for this in fact available or accessible? For the US stockpiles, the answers to such questions are, for the most part, classified, though at a relatively low level. For the French and Soviet stockpiles, the open literature offers nothing.

WTO countries

A West German authority has recently stated that "the Soviet Union ... has available a potential of 200 000 to 700 000 tons of theater chemical weapons" (Rühle, 1977, 1978). Two years previously the Chairman of the US Joint Chiefs of Staff had said that "it is not possible with any reasonable degree of assurance to predict or estimate the size of the USSR's CW agent stockpile. Other evidence however, reflects a requirement for a sizeable stockpile both in bulk agent and filled munitions. The USSR's stockpile is probably more than adequate to meet its minimum requirements" (Brown, 1975 *a*).

It would seem that this last statement is based on two quite different approaches to stockpile estimation. One starts from observations of capability: the size of putative production facilities, for example, or the roof area of possible storage sheds. A "reasonable degree of assurance" might well be difficult to achieve with this approach, though the available methodologies are said to have improved since 1975, and the technique lends itself to continuous refinement as additional data are acquired. The other approach starts from indications of intent: from, for example, internal doctrinal studies or manuals relating to the conduct or planning of chemical operations from which some idea of agent or munitions requirements can be inferred.

The second approach is apparently the basis for the numerous estimates of Soviet stockpile that have appeared in the open literature over the years which are expressed as percentages of total stockage of munitions, sometimes for particular categories of weapon system. The oldest of these, which persisted for much of the 1950s and 1960s, was that one-sixth of the total Soviet munitions stockpile was chemical (e.g., Trudeau, 1960; Gadsby, 1968). Later versions appearing in the Western press have included the following, the contradictions between them presumably reflecting an uncertainty which they do not, however, remark:

— 25 per cent of the total Soviet stockpile of artillery shell and 45 per cent of the rockets, or 35 per cent of the total stockage of bombs, rockets and shell (*Soldat und Technik*, 1968; Deutsche Welle, 1968);
— one-third of the Soviet anti-personnel land-mines stored in the GDR (Deutsche Welle, 1968);
— 15 per cent of Soviet munitions stocks in the GDR (*Der Spiegel*, 1969);
— more than one-third of the available artillery and air-force munitions (*Allgemeine Schweizerische Militärzeitschrift*, 1970);
— nearly one-third of all ammunition, rocket launchers, rockets and bombs (*Soldat und Technik*, 1971); and
— of Soviet munitions in the GDR, "a quarter of the aircraft bombs and numerous rocket warheads" (*General-Anzeiger*, 1976).

A more recent expression is that, "of the stockpiles maintained for individual weapons systems, 5 to 30 percent consist of chemical munitions" (Rühle, 1977, 1978).

With the exception of the *Der Spiegel* article, which spoke of "nerve gas" munitions, none of these estimates specified the category of CW agent in the munitions to which they related. So far as is known from open sources, the USSR has not yet been observed discarding significant quantities of its World War II chemical stocks. A 1960 US publication stated that "the USSR probably has more mustard than any other agent", those which it listed being adamsite, hydrogen cyanide and tabun (*Armed Forces Chemical Journal*, 1960). Other older types of lethal chemical agent reportedly in the post-World War II Soviet stockpile include chloropicrin, phosgene and diphosgene (*Soldat und Technik*, 1970). Open reports of soman being stockpiled did not begin to appear until the late 1960s (e.g., Ganas, 1969), by which time there were also reports of a novel CW agent which the USSR was said to have designated VR-55 (*Soldat und Technik*, 1970). VR-55 is now thought to be a thickened form of soman. Possibly it was this to which a recently retired Director of the US Defense Intelligence Agency was referring when he spoke in a 1976 press interview (Anderson, 1976) of a new Soviet nerve gas which he called "V gas"— or, in the version of Menaul (1978), "B gas".

At one time or another, Western sources have referred to the existence of chemical munitions for almost all Soviet weapons systems that are in principle suited to chemical warfare. These references are collated in table 1.

Table 1. US and Soviet chemical weapons: a comparison

		Reported chemical weapons of the USSR[a]				Chemical weapons of the USA[b]			
			Data on three indicators of effectiveness or utility						
			Max. range (km)	Mass (kg) of CW agent that can be placed on target per fire-unit[c]				Max. range (km)	
Category of weapon	Ref. to notes below	Type		During 15 min	During 15 sec	During 15 sec	During 15 min		Type
Land mines	*d*	KhF	0	n.a.	n.a.	n.a.	n.a.	0	2-gal VX
						n.a.	n.a.	0	1-gal mustard
Mortars	*e*	120-mm	5.7			70	1 100	5.5	107-mm mustard
Cannon	*f*	122-mm how	24			80	1 700	11	105-mm how mustard or GB
		152-mm gun-how	17	1 000–2 500	50–150	100	2 300	18	155-mm how mustard, GB or VX
		180-mm gun-how	32			90	900	21	203-mm how GB or VX
Multi-launch rocket systems	*g*	122-mm 40-tube	20	7 000–12 000	3 000–6 000	700	700	11	115-mm 45-tube VX or GB
		250-mm 6-tube	30	5 000–8 000	5 000–8 000				
Artillery rockets	*h*	FROG-7B	70	*c.* 600	*c.* 600	(All CW warheads declared "surplus" in 1973)			Honest John, GB
Guided missiles, ballistic and cruise	*i*	Scud-B	280	*c.* 400	*c.* 400	(CW warhead standardized but apparently not procured)			Sergeant, GB or VX
		Shaddock	450	*c.* 600	*c.* 600				
Tactical aircraft	*j*	(No details have been published on the CW bombs and spraytanks that are reported)				2 500	2 500	600	A4 w/500-lb bombs, GB
						1 900	1 900	650	F4 w/750-lb bombs, GB
						1 200	1 200	650	F4 w/160-gal tanks, VX
Naval ordnance	*k*	(No details have been published on the Soviet naval CW weapons that are reported)				(CW projectiles were apparently standardized but never procured)			5-in/38 gun, GB or VX
									5-in/54 gun, GB or VX
									5-in 48-tube MLRS, GB

Conventions and abbreviations: "n.a." stands for "not applicable"; "w/" stands for "with"; "how" stands for "howitzer"; and "MLRS" stands for "multi-launch rocket system".

Notes and sources:

[a] According to the Western sources cited in notes *d–k* below; no references to Soviet chemical weapons exist in the open Soviet literature. The weapons listed in the table are ones in Soviet service for which Western reports of the availability of chemical ammunition are most frequent; others are noted below. The figures given for maximum range refer, where possible, to the latest versions. The information from which the figures on weapon agent-delivery capability were estimated—i.e., rate of fire, munition weight and basis of issue—came from Western publications that are often at variance with one another; evidently the data which they purvey are often rough estimates.

[b] According to official US sources, primarily Army field manual FM 3–10. The weapons listed here are ones for which nerve-gas or mustard-gas ammunition is currently available. Data on range and rate-of-fire are for the latest weapons capable of firing current supplies of chemical ammunition. Where more than one type of chemical munition is available for a weapon, the figures given on agent-delivery capability refer to the munition of highest capacity.

[c] The agent-delivery capabilities given for the weapons are estimated, not on a per-weapon basis, but according to the fire units in which the weapons would, on current US or Soviet practice, most probably be used. Particulars are as follows:

Weapon		Fire unit and number of weapons in it		Remarks
USSR	120-mm mortar	Battery	6	–
USA	107-mm mortar	Battery	4	This, the only US mortar for which CW rounds (mustard only) are available, is shortly to be phased out (Foss, 1978).
USSR	122- and 152-mm cannon	Battalion	18	–
USA	105- and 155-mm cannon	Battalion	18	–
USSR	180-mm	(Battalion)	12 (assumed)	Basis-of-issue data unavailable.
USA	203-mm	Battalion	12	–
USSR	122-mm MLRS	Battalion	18	Though some Western sources refer to 24 launchers per battalion.
USSR	250-mm MLRS	Battalion	18	–
USA	115-mm MLRS	(Battalion)	3	The original, 1961, basis of issue was 36 M91 launchers per battalion. Now being phased out, the weapon is carried as a Class II item in the service battery of 105-mm howitzer direct support battalions. Though obsolescent, its availability with both GB and VX M55 rockets is still noted in current Army manuals (e.g., US Dept of the Army, 1977*d*).
USSR	FROG-7B	Battalion	4	–
USSR	Scud-B	(Battalion)	2 (assumed)	Basis-of-issue data unavailable.
USSR	Shaddock	(Battalion)	2 (assumed)	Basis-of-issue data unavailable.
USA	500-lb bomb	A-4M Skyhawk II	16 (assumed)	Figure refers to the high-capacity Mk116 Weteye bomb, which has been qualified for the A-4, rather than the Mk94.
USA	750-lb	F-4C Phantom II	19	Data on loading of MC-1 bombs and TMU-28 spraytanks from USAF Phantom flight manual.
		F-4C Phantom II	2	

The sizes of the targets against which the listed chemical weapons might be used would depend on several sets of factors, one set being the dosage-distribution characteristics of the munition and the dosage-response (toxicity) characteristics of the CW agent distributed by the munition. The former will be highly dependent upon prevailing weather conditions; the latter, upon the degree of anti-chemical protection of the target population. Both dependencies are reflected in the munitions-expenditure tables available for particular munitions and delivery systems. The following figures are calculated from such tables: they indicate the approximate mass of agent that must be distributed over the target per unit of target surface-area in order to achieve around 50 per cent casualties among unprotected personnel in the open under weather conditions that are neither especially favourable nor unfavourable for chemical warfare. Depending on ambient temperature, wind vectors, terrain and whether contact or respiratory effects are sought, actual requirements will vary within an order of magnitude of these figures:

Agent:	Hydrogen cyanide	Mustard gas	Sarin nerve gas	Thickened soman	VX nerve gas
Density:	250 kg/ha	150 kg/ha	10 kg/ha	5 kg/ha	3 kg/ha
Source:	World War II US–UK tables for 1 000-lb bomb (Latimer, 1946)	US Dept of Army (1971 *b*) tables for 105-, 107- and 155-mm HD rounds	US Dept of Army (1971 *b*) tables for 105-, 155- and 203-mm GB rounds	Extrapolation of VX data (*q.v.*)	US Army data on 155-mm rockets and 2-gal mines in Ganas (1969)

[d] The KhF is described as a "bounding chemical mine" (US Dept of the Army, 1975). A French official has referred to Soviet mustard-lewisite land-mines (Ganas, 1969). Forward deployment of Soviet chemical land-mines, in the GDR, is reported (Deutsche Welle, 1969).

[e] Official US sources (e.g., US Department of the Army, 1976 *a*) refer to chemical ammunition for Soviet mortars, including the 240-mm heavy mortar (US Dept of the Army, 1975) which, however, is no longer in service with Soviet units. Finan (1974) records a chemical capability for the 120-mm mortars, which are the standard regimental mortars. Forward deployment of chemical mortar ammunition is reported (*Newsweek*, 1976).

[f] Official US sources (e.g., US Dept of the Army, 1975, 1976 *a*) refer to chemical ammunition for Soviet cannon artillery. Finan (1974) records a chemical capability for the M1938 and D-30 towed 122-mm howitzers, as does Gunston (1976) for the D-74; these are standard divisional-artillery weapons. They are being replaced, it is believed (Pretty, 1977), by the M1974 self-propelled 122-mm howitzer, which mounts the same tube as the D-30. As regards heavier artillery, Finan (1974) records a chemical capability for the M1938 (M10) 152-mm howitzer, as does Pretty (1977) for the M55 152-mm towed gun-howitzer; these are also standard divisional-artillery weapons. Chemical ammunition is reported for long-range artillery systems organic to formations above division, in particular for the 130-mm M-46 towed field gun (Gunston, 1976) and for the recently introduced S-23 180-mm towed gun-howitzer (US Dept of the Army, 1975; Pretty, 1977).

[g] Multi-launch rocket systems (MLRS) are well suited to surprise chemical attacks on area targets. Large numbers of several different types are in service with Soviet and other WTO forces. Both nerve-gas and hydrogen-cyanide rockets are said to be available (Ganas, 1969). Dehn *et al.* (1967: 93) refer to hydrogen-cyanide rockets for MLRS. The self-propelled 122-mm BM-21 MLRS, a standard divisional artillery weapon introduced in 1964, is said to be one of the principal components of the Soviet offensive CW capability (Finan, 1974; Donnelly, 1976; Pretty, 1977; Accasto, 1978; *et al.*). Other particular MLRS for which chemical rockets are reportedly available include the much older 140-mm BM-14 16-tube system (Finan, 1974) which is no longer in front-line service (Pretty, 1977), apparently having been replaced by the 17-tube version introduced in 1959 or the new 16-tube M-1965 version (Hoffman, 1977), and the 250-mm BM-25 system (Finan, 1974) which is said to be a divisional-

artillery weapon. A new large-calibre MLRS—said to be a 10–12 tube *c.* 240-mm system—is reportedly entering Soviet service (*International Defense Review*, 1978 *c*; Dick, 1979), but there have been no reports yet of chemical ammunition for it.

[h] Battalions of heavy surface-to-surface free-flight rockets (NATO: FROG; USSR: T-5) have been organic to Soviet and other WTO divisions since the late 1950s. Of the several successive versions of these nuclear/non-nuclear dual-capable weapons, chemical warheads have been reported for all except, apparently, the first (Gatland, 1976: FROG-2, -3, -4, -5 and -7; *Neue Zürcher Zeitung*, 1975: FROG-3; Finan, 1974: FROG-4 and -7; US Dept of the Army, 1975: FROG-3 and -7). For FROG-7 (USSR: Luna), the chemical warhead is said to be somewhat longer but lighter than the nuclear warhead (Wood and Pengelley, 1977). According to the US Army, the warheads are of the bulk filled (as opposed to bomblet) type, designed to open high over a target area to release a rain of persistent nerve gas (Meyer *et al.*, 1978). A French official has referred to hydrogen cyanide as a FROG payload (Ganas, 1969). A chemical capability "seems probable" according to *International Defense Review* (1979 *a*) for the SS-21 missiles which are said to be replacing FROGs, and whose deployment with Soviet forces in the GDR is now reported (*International Defense Review*, 1979 *b*), though apparently erroneously (*Ground Defence International*, 1979).

[i] What the USSR calls "operational-tactical" missile systems, nuclear/non-nuclear dual-capable, are organic in regiment strength in Soviet fronts and armies, the most widely deployed weapons in this category being the mobile Scud (NATO; USSR: T-7) ballistic missiles. Scaleboard (US: SS-12), a somewhat larger and longer-range version of late-model Scuds, and possibly also the Shaddock ground-launched cruise missile, may be similarly available, though in smaller numbers. Chemical warheads are reported for both versions of Scud (Finan, 1974, and Gatland, 1976: Scud-A; Middleton, 1972: Scud-B; US Dept of the Army, 1975: Scud-A and -B); they are said to be similar in design and operation to those available for FROGs (Ganas, 1969; Meyer *et al.*, 1978). For Scaleboard, chemical warheads are said to be a possibility (Marriott, 1977). The 1979 military posture statement by the Chairman of the US Joint Chiefs of Staff says that "possibly cruise missiles" are included among Soviet chemical weapons (G. S. Brown, 1978). No reports yet seem to have been published of chemical warheads for the Scud and Scaleboard replacements, SS-23 and SS-22 respectively.

[j] The US and West European literature contains several references to Soviet chemical air munitions, but little detailed information is available. The Chief Chemical Officer of the US Army stated in 1959 that "converted . . . incendiary bombs and rotational scattering aircraft bombs are used for chemical warfare purposes" by the USSR (Stubbs, 1959). A French official has referred to massive bombs charged with hydrogen cyanide or soman and to fragmentation bombs containing mustard gas or incapacitating agents (Ganas, 1969). The elderly Il-28 Beagle strike aircraft has been portrayed as a chemical-munitions delivery system (Malooley, 1974), as indeed those of the Egyptian Air force appear to have been used during the Yemeni Civil War of the mid-1960s (SIPRI, 1971 *a*: 336–41). During World War II, German forces captured stocks of Soviet air chemical munitions, including: the AK-2 downward-ejection dispenser munition containing about 240 one-kilogram frangible bomblets filled with mustard-lewisite mixture (*Chemical Corps Journal*, 1949); several sizes of bomb, such as the CHAB 500 charged with 170–180 kg of phosgene; and several sizes of aircraft spraytank, such as the BATT (VAP) 1000 dispensing about half a ton of hydrogen cyanide (Mills and Harris, 1945).

[k] A chemical capability is available for Soviet naval forces, according to official US sources (e.g., Brown, 1975 *a*); little further information is available.

Western commentators lay particular stress on hydrogen-cyanide and nerve-gas warheads for the various rocket-artillery systems (e.g., Donnelly, 1976; Hauger, 1977; Watson, 1978, Meyer *et al.*, 1978).

As to deployment, the German Federal Government (1970) has stated that Soviet forces in the GDR, in Czechoslovakia and in Poland maintain large stocks of chemical weapons. There are US press reports of further stocks in western parts of the Soviet Union (Middleton, 1976) and in the vicinity of front-line divisions and air regiments deployed along the Chinese border (Middleton, 1978); in both respects, US Government officials (e.g., Lennon, 1978) have made similar statements. An authoritative, but in this case most unreliable, British source (Erickson, 1976) states that at least 40 Soviet depots contain CW and BW agents. A US press report of low reliability states that Soviet front-line Rifle Regiments in east-central Europe have chemical mortar and artillery ammunition stored in the same bunkers as conventional ammunition (*Newsweek*, 1976).

It would be reasonable to assume that, as in the case of US chemical munitions, any Soviet stocks deployed on foreign territory would remain solely under Soviet control. However, a recent article in a reputable Swiss military journal (*Allgemeine Schweizerische Militärzeitschrift*, 1978) includes the statement that the GDR *Volksarmee* is being equipped with Soviet chemical weapons; the British Government has stated, however, that this article contains "errors" (Peart, 1979). The journal has since published a new version of the report (*Allgemeine Schweizerische Militärzeitschrift*, 1979).

NATO countries

Current NATO policy, as understood by the United States, directs that NATO forces in all regions have a capability to employ lethal agents in retaliation (Rogers, 1978). It is not clear from the public record just how far each of the NATO allies has advanced in acquiring such a capability, particularly since the Federal Republic of Germany has renounced chemical weapons in perpetuity, and Britain and Canada at least for the time being. However, in 1975, the US Congress was informed by the Chairman of the Joint Chiefs of Staff that "our NATO allies have weapons capable of delivering chemical munitions and do develop requirements which are submitted to NATO headquarters on the assumption that the United States would provide CW munitions for retaliatory purposes" (Brown, 1975*b*). Since then, the question has occasionally been raised in public (e.g., *Nature*, 1979), and no doubt also in private, of whether other NATO countries that are not precluded from doing so by international treaty should begin producing chemical munitions of their own instead of relying on US stocks. The case of Britain is of particular relevance, but the present British Government has just followed its predecessors in stating that there are no plans to equip British forces with a retaliatory chemical capability (Hayhoe, 1979).

Estimates that may be made on the basis of unclassified information[12] about current US stocks suggest that somewhere in the region of 38 000 tonnes of CW casualty anti-personnel agents are available, slightly less than half being sarin and VX nerve gases (in a ratio of about 3 to 1) and slightly more than half comprising three forms of mustard gas (Levinstein mustard, distilled mustard and HT mixture) remaining from World War II. By decision of President Nixon in February 1970, the USA renounced production and stockpiling of "toxins produced either by bacteriological or biological processes or by chemical synthesis" (US National Security Council, 1970) and the Army's stocks of PG and TZ—both of them toxins that had previously been categorized as chemical agents—were thereafter destroyed. The stocks of BZ and BZ munitions, by now apparently having been declared obsolete (Richards, 1979), would seem to be insignificantly small: only about 50 tonnes of BZ (a psychotropic glycollate) were produced, during 1963–64 (Rosenblatt *et al*., 1977). Almost two-thirds of the tonnage of mustard gas and about one-quarter of the tonnage of nerve gas appears to be held in bulk form, in 1-ton storage containers. The remainder is held in munitions, of which the total supply seems to be in the range 150 000 to 200 000 tonnes. If all of the bulk agent were to be filled into munitions, the total supply would increase to around 400 000 tonnes.

Nearly three-quarters of the current filled-munitions tonnage appears to comprise nerve-gas munitions, of which 30 or so varieties have been approved for the operational inventories of US armed forces over the past 30 years, though only about half of this number seems in fact to have been procured in quantity. Some, like the M34 1 000-lb sarin cluster bombs, are now obsolete and have been discarded, as also, though for other reasons, have the stocks of sarin warheads for Honest John rockets. Remaining in the stockpiles are around three million sarin and VX artillery projectiles; of the order of thousands—possibly tens of thousands—of sarin-filled aircraft bombs, these, like the artillery shell, having been procured intermittently over the period 1953–69; some thousands of 115-mm rockets procured during 1960–63 (mostly VX, probably; at least 118 000 of the sarin-filled versions have been discarded since then, or scheduled for destruction); some hundreds of thousands of VX land-mines procured during 1961–66; and about 1 500 2 000-lb VX-filled aircraft spraytanks procured during 1965–68. In addition to these nerve-gas munitions, there seem to be between 1 and 4 million mustard-gas projectiles, which must by now be in an uncertain state of reliability. Further details are given in table 1.

An undisclosed, but large, fraction of the total US stocks of nerve-gas and mustard-gas munitions is categorized by the US Defense Department as unserviceable, obsolescent or obsolete (Olenchuk, 1975; Cooksey, 1975 *c*), though a Congressional investigation has called some of this assessment into question (US Comptroller General, 1977). The 1980 defence budget, currently before the Congress, contains a $21.8 million line-item for "renovation of existing chemical munitions inventories"; $4.6 million had been

Table 2. Locations of the major US stocks of mustard and nerve gases

Location	Estimated stocks[a] as percentage of total	Information on inventory
Tooele Army Depot, Tooele, Utah	39–43	Nerve and mustard; includes stocks of almost every type of filled (H, HD, HT, GB or VX) munition currently operational; eight $\frac{1}{2}$-mile rows of 1-ton bulk storage containers reported.
Pine Bluff Arsenal, Pine Bluff, Arkansas	12–13	Nerve and mustard, some or all of the latter in 1-ton bulk storage containers.
Umatilla Army Depot, Hermiston, Oregon	12–13	Nerve and, probably, mustard, some in bulk storage containers; includes all of the CONUS stocks of Navy Mk94 500-lb GB bombs.
Anniston Army Depot, Anniston, Alabama	9–10	Nerve and mustard, some or all of the latter in bulk storage containers.
Newport Army Ammunition Plant, Newport, Indiana	3–6	VX only.
Pueblo Army Depot, Pueblo, Colorado	3–6	Mustard only.
Johnston Island, Pacific Ocean	3–6	About 12 000 tonnes of nerve and mustard munitions in all (15 per cent VX, 63 per cent GB, 22 per cent mustard); includes Mk94 500-lb GB bombs, 115-mm M55 nerve-gas rockets, and howitzer projectiles.
Edgewood Arsenal, Edgewood, Maryland	3–5	Mustard only, some or all in bulk storage containers.
FR Germany[b]	2–4	Nerve gas only.
Bluegrass Army Depot, Lexington, Kentucky	1	Nerve and perhaps mustard gas.
Rocky Mountain Arsenal, Denver, Colorado	<1	Prior to 1973, the second largest chemical-weapons depot; after the recently concluded demilitarization operations there, the only sizeable operational stocks remaining seem to be about 900 Mk116 (Weteye) 500-lb bombs, the entire supply.

Notes:

RDT & E quantities of chemical agents or munitions are stored, also, at Redstone Arsenal, Alabama, and Dugway Proving Ground, Utah.

[a] In terms of CW agent, whether stored in bulk containers or in munitions, the total apparently being of the order of 40 000 tons. Estimated by the methods used in Robinson (1975 *a*: 35–37), refined in the light of subsequent information. A recent major open source comprises testimony before the House Military Construction Appropriations Subcommittee by US Defense Department witnesses on the security of nuclear and chemical weapon storage sites: US Dept of Defense (1978).

[b] At one location only, it is officially stated (ibid.). This is presumably one of the four US Army chemical depots that have been mentioned in the press: at Hanau, Massweiler, Viernheim and Mannheim.

programmed for the same purpose in the 1979 budget (H. Brown, 1979: 176). It is not yet clear from the public record what exactly this $26.4 million is for, but it appears to include the "rehabilitation" (Rogers, 1979 *a*) of "unserviceable chemical munitions" (US Arms Control & Disarmament Agency, 1979).

Only a small fraction of US supplies of chemical munitions has been located outside the United States, apparently about one-eighth, either on Johnston Island in mid-Pacific or in FR Germany. Particulars of the stockpiles inside and outside the United States are given in table 2. The stocks in FR Germany, solely under US control, contain perhaps 500 to 1 000 tonnes of nerve gas, most of it in ground munitions. This supply is thought to be similar in size to that of France.

The US Army has about 500 000 tons of munitions of all types stockpiled in Europe (Center for Defense Information, 1978). The chemical munitions in FR Germany—of the order of 10 000 tonnes—therefore constitute around 2 per cent of the total. If all of the filled chemical munitions at present stored in the United States were to be moved into Europe, the proportion would rise towards 30 per cent. Contingency plans for just such a forward deployment action have been developed by the US Joint Chiefs of Staff (Rogers, 1978). Serious political obstacles would, however, confront their execution, though, should East–West tension increase seriously, FR Germany or the UK, for example, might become more willing than they are now to provide additional depot facilities for US chemical munitions. Geographical obstacles would remain. It has been estimated that the airlifting of three million artillery rounds from the United States into Europe would take two weeks or more to accomplish even if most of the currently available US military air transport capacity were used for the purpose (*Arms Control Today*, 1979).

VII. Conclusions

In terms of production capacity, anti-chemical protection and total stocks of CW agents, the overall impression conveyed by the open literature is that the WTO and NATO have roughly comparable capabilities for chemical warfare.

This is not, however, the opinion that is dominant in the West. Statements on the matter by leading US defence officials during 1978 (H. Brown, 1978 *a*; G. S. Brown, 1978; Haig, 1978; Alexander, 1978 *b*; Perry, 1978: 5638), as in earlier years, described Soviet/WTO chemical warfare capabilities as substantially superior to those of NATO, the official view being, moreover, that the capabilities had been increasing for some time and were continuing to do so (Hoffman and Rogers, 1977: 652; Miller, 1977; Watson, 1978; Davis, 1978 *b*). Though the 1979 annual reports to the US Congress by the Chairman of the Joint Chiefs of Staff (Jones, 1979) and the Secretary of Defense (H. Brown, 1979) are much more muted in this respect, the perception of inferiority appears to have become widely, even unquestionedly, accepted within US and West European CW policy-shaping circles. Contributory, perhaps, is the publicity that has been given to a number of unofficial, undocumented, but purportedly well-informed, commentaries on the matter (e.g., Menaul, 1978; Hoeber and Douglass, 1978).

There are several possible explanations for the contradiction between these perceptions and the picture that emerges from the present review. One is that the open literature is inadequate: that its coverage of WTO capabilities is too incomplete to allow any such comparisons, and that it is misleading on the operational readiness of NATO capabilities, both protective and offensive.

A second explanation is that the secret literature, like the open literature, is also inadequate: in other words, that the improvements in Western CW intelligence capacity (stimulated by the opening of intergovernmental CW disarmament talks in the late 1960s) have not in fact been good enough to admit detailed capability comparisons. The possibility always exists in a field as obscure and disquieting as CW that the resources committed to the gathering and assessment of pertinent intelligence will be inadequate, and that there will be a disinclination on the part of decision makers to question the results. This is a situation in which checks and balances against bias, prejudice and vested interest may be weak; where exaggeration may be portrayed as moderation; and where the fear may come to be perceived and then advertised as the reality. An outsider, such as the present author, can of course offer no informed judgement. All he can do is to remark consistencies, inconsistencies and transiences in the open historical record, thereby creating a modest background against which can be measured those more extravagant claims that are facilitated by the shortness of bureaucratic memory.

A third explanation is that, while on some criteria of capability NATO and WTO may indeed be evenly matched, there are other more important criteria on which they are not. The matter of asymmetric forward deployment of chemical munitions, for example, certainly weighs heavily with those Western commentators who believe in the possibility of in-kind CW deterrence. "Authoritative NATO sources" have been quoted as saying that the allied chemical-weapons capability is only a quarter of what it should be (Farr, 1979*b*). Even though Soviet forward stocks in Europe are as liable to deterioration and obsolescence as those of the United States, their refurbishment and replenishment is much easier, both politically and geographically; and replenishment has been reported to have been taking place over the past 10 years[13] (*Soldat und Technik*, 1968, 1971); *Allgemeine Schweizerische Militärzeitschrift*, 1970, 1978; *Newsweek*, 1976; Middleton, 1976; *International Defense Review*, 1978 *b*, 1979 *a*; Weinstein, 1978; *Economist Foreign Report*, 1978). Degree of assimilation of CW capability may be another such criterion (Robinson, 1977): some Western commentators believe the Soviet command to have fully integrated CW into the structure, training and equipment of all branches of the Soviet armed forces, including ground, air and naval units (e.g., *Defense and Foreign Affairs Daily*, 1978; Henry, 1979). This is a matter that has not been explored directly in the present review.[14] It is clear, given the wide variation in the national CW policies of NATO member states, that chemical weapons are poorly assimilated into overall NATO doctrine and force structure; but the same cannot be said of anti-chemical protection.

A final possibility is that Western commentators are responding to what is in fact a transient asymmetry in NATO and WTO capabilities, one which has significance only in the short term. Several of the NATO armed services have for some time been embarked upon major programmes of re-equipment, indoctrination and training in anti-chemical protection which are now coming to fruition, thereby outmoding those perceptions of inferiority which stimulated the programmes in the first place. But, as regards the future course of intergovernmental CW arms-limitation negotiation, the danger is that these perceptions may nonetheless have taken root: that, unless confidence can be built, they may flourish to such an extent that Western requirements for treaty-verification procedures will remain unfulfillably demanding.

Notes

[1] Where unattributed opinions are presented in this paper, they are those of the author and should not be construed as those of the institutions with which he is affiliated or otherwise associated. The author wishes to acknowledge with gratitude support received during the preparation of this review from the Joseph Rowntree Charitable Trust.

[2] A somewhat ambiguous reference to Soviet chemical weapons is to be found in S. Talbott (editor and translator), *Kruschev Remembers* (Boston: Little, Brown, 1971) on page 518:

> Our army would be in a miserable situation if our enemy were to use chemical and bacteriological weapons against us and we didn't have any of our own. As long as two opposing systems exist, we will be obliged to keep all possible means of warfare stockpiled. I'm emphasizing this because I want my belief in the importance of vigilance and effective deterrence against imperialist aggression to be clearly understood.

[3] A fragment of, allegedly, the "Limited War" section of JSCP-62, Joint Strategic Capabilities Plan, itself a part of the US Joint Chiefs of Staff War Plan JCS 61-62, surfaced some while ago; if genuine, it provides an instructive picture of how CBW had come to be viewed by the Joint Chiefs in the early 1960s:

> f. Potential of chemical–biological weapons must be exploited to the maximum possible extent. This objective includes use of chemical–biological weapons whenever tactically appropriate without regarding them as any special form of warfare. The uninhibited use of disabling but nonlethal agents may provide the initial breakthrough.

The context of this purported extract remains unpublished. Overall CW policy at that time required Presidential authorization prior to release of CBW weapons to field commands. Plans and doctrine are, of course, very different from capability; and in this case it is clear that the latter was never acquired to an extent commensurate with the former. Even under present US CW policy, the Defense Department has on several recent occasions told the Congress that the mismatch between chemical-weapons posture and doctrine is considerable.

[4] The 1974–79 Army, Navy, Marine Corps and Air Force reviews of chemical-warfare posture appear to have been stimulated by two principal factors and promoted by a

third: the discovery of large quantities of modern Soviet anti-chemical equipment among Israeli captured matériel after the 1973 Yom Kippur War; Congressional opposition to the funding of chemical weapons programmes; and disengagement from the Vietnam War. At the CW-specialist level within the Service Departments there had, no doubt, long been awareness of deficiencies in US anti-chemical protection but, in terms of bureaucratic influence, the voices of the people most concerned would not have been strong. A devastating critique of the state of the Army's CW protective posture was published by a Chemical Corps officer soon after the matter had begun to attract top-level attention (Templeton, 1975). Particulars of the subsequent Army, Air Force, Navy and Marine Corps CW-defence upgrade activities are to be found in the military journals (e.g., Henry, 1979; *Marine Corps Gazette*, 1979; *National Defense*, 1979) and in the following Congressional testimony: Augustine and Cooksey (1975); Blanton (1976); G. S. Brown (1975*b*, 1975*c*, 1976); H. Brown (1979: 176); J. B. Currie (1978); M. R. Currie (1975); Feir (1976); Feir, Greiner and Jones (1977); Hoffman and Kerwin (1976); Hoffman and Rogers (1977: 690); Keith (1978); Kjellstrom (1976, 1977); Meyer (1978*b*); Miller (1976, 1977); Olenchuk (1974, 1975); Rogers (1979*b*); Schlesinger (1975); US Dept of Defense (1976*a*, 1976*b*, 1977); Watson (1978); Weyand (1975). Especially notable is the increased attention now being given to CW training both at the unit level and in major exercises—as, recently, in REFORGER and Gallant Crew (Alexander, 1978*a*). Sufficient quantities of protective equipments are being procured not only for war reserve stocks but also to allow for ample consumption during training and exercises. A new NBC training facility has been established at Redstone Arsenal, Huntsville, Alabama, in addition to the Ordnance and Chemical Center and School at Aberdeen Proving Ground, Maryland (Miller, 1977; Watson, 1978).

[5] Except, at higher ambient temperatures, under conditions of very strenuous work (US Dept of the Army, 1977*c*). But data on tolerance-times versus relative humidity, for example, have not yet been disclosed.

[6] These matters are discussed and illustrated in detail in appendix 5, unpublished, of volume 2 of SIPRI's *The Problem of Chemical and Biological Warfare* (SIPRI, 1973: 276).

[7] The source given for this information, which purports to refer to the period up to 1966, is Professor A.-H. Frucht, formerly Director of the Institut für Arbeitsphysiologie in Berlin-Lichtenberg and now living in the West. In 1968 he had been convicted of espionage and sentenced to life imprisonment from which he was released nine years later as part of a complex international exchange.

[8] Pozdnyakov also speaks of a pre-World War II Soviet CW test facility at Gorokhovetsky Camp near Gorki; and the "Kuzminki Polygon", near Moscow, to which he also refers, could be the test range noted in the 1949 US intelligence estimate as the "Moscow Gas Polygon", 20 km outside Moscow, described as having been the main Soviet CW experimental field until 1932, when it was transferred to the more spacious accommodation at Shikhany. The Moscow Gas Polygon was said to have out-stations at Pskov, near Leningrad, and Nikolsk, near Vladivostok. Kuzminki is now an inner residential suburb of Moscow.

[9] For further details, though now rather outdated, on these and other NATO CW R & D establishments, see SIPRI (1973: 202–24). This publication also touches on the particulars of university and industry involvement in some of the R & D programmes.

[10] The route to sarin chosen for the US production programme was the DMHP process, the choice reflecting the considerable amount of development work done on

the process in wartime Germany. Details of this and other routes to sarin and other nerve gases are given in appendix 4, unpublished, of volume 2 of SIPRI's *The Problem of Chemical and Biological Warfare* (SIPRI, 1973: 53).

[11] Further back in development are several other binary nerve-gas munitions that might in due course also be provided for in the Pine Bluff facility:

— a warhead for the Lance battlefield-support guided missile;
— a warhead for the 70-mm air-to-ground rocket;
— a warhead for the developmental 227-mm (Multi-Launch) General Support Rocket System;
— a projectile for the Light Weight Company Mortar, 60 mm or 81 mm;
— an extended-range, sub-calibre, fin-stabilized projectile for 203-mm artillery; and
— a 155-mm projectile disseminating an intermediate-volatility agent, such as EA 5365.

[12] The derivation of most of these estimates of US chemical-weapons stocks, and literature references to the published data used, are set out elsewhere (Robinson, 1975 *a*, 1977; SIPRI, 1975), though here the estimates are somewhat revised in the light of new information, especially a recent press report quoting "Congressional sources" on certain stockpile details (Albright, 1978) and Defense Department congressional testimony on the security of chemical weapons storage sites (US Dept of Defense, 1978). One basic assumption used in deriving the tonnage estimates for munitions is that, on average, about 10 per cent of the mass of chemical munitions is contributed by their payload of CW agent. US CW artillery shell contain rather less agent than this, and such things as aircraft bombs and spraytanks considerably more; but the US CW stockpile is known to be heavily weighted towards artillery capability.

[13] In relation to what is suggested on pages 12–13 about action–reaction dynamics in NATO and WTO chemical armament processes, it may be observed that, if the reports of a Soviet CW build-up over the past 10 years are correct, the initiating decisions in Moscow would have been taken in the early 1960s—at the time, in other words, of the last high-point in US chemical-weapons procurement.

[14] One factor in particular which may both affect and reflect degree of assimilation is the level at which authority to use chemical weapons rests. Since the early part of World War II, US policy and doctrine on chemical weapons has always required Presidential release prior to delegation of employment authority to military commands. One effect of such tight political control over the weapons—necessitated, apart from anything else, by the country's no-first-use obligations under the 1925 Geneva Protocol—has been to ensure their containment within a special category, and thus to isolate them from routine military planning and training activities: the military cannot be expected to devote full attention, unreluctantly, to weapons they may never be permitted to use. In this as in several other respects there is a strong parallel between chemical and theatre-nuclear weapons. Exactly where Soviet CW release authority resides is a matter of some dispute in the West. Several years ago, *Allgemeine Schweizerische Militärzeitschrift* (1970) reported that authority had been delegated to divisional commanders, which was the first time that such a remarkable claim had been made in the pages of a reputable publication. Similar statements, no less deficient in authentication, have since been made by several Western commentators (e.g., *Soldat und Technik*, 1971; Marriott, 1977; *Defense & Foreign Affairs Daily*, 1978), including at least one authoritative analyst of Soviet military affairs (Erickson, 1977) who had previously rejected the notion. The view of the US Joint Chiefs of Staff on the matter, as

of early 1978, is this: "[Soviet chemical-weapons] doctrine... requires initial release authority to Politburo level, then delegation to 'Front' commands as required" (G. S. Brown, 1978). While it is true that US officials believe there to have been a change in Soviet CW release doctrine in recent years, the opinion prevailing among them appears to be that the change is a decoupling of chemical weapons from nuclear weapons (e.g., Lennon, 1978). The net result of this would be to bring WTO practice into line with that of NATO.

References

Accasto, M., 1978. Artillery rocket-launchers. *Armies & Weapons*, No. 41 (February–March), pp. 37–44.

Agursky, M., 1976. The Research Institute of Machine-Building Technology. *Soviet Institutions Series* (Hebrew University of Jerusalem: Soviet and East European Research Center), Paper No. 8 (in Russian, English summary).

Albright, J., 1978. Anniston: safest nerve gas arsenal? *Journal and Constitution* (Atlanta), 12 February.

Aleksandrov, V. N., 1969. *Toxic Agents*. Moscow: Voyenizdat (in Russian).

Alexander, C. L., Jr (US Secretary of the Army), 1978 *a*. Prepared statement on the FY 1979 Army budget. In: US Congress, Senate Armed Services Committee (1978), part 2, p. 678.

Alexander, C. L., Jr (US Secretary of the Army), 1978 *b*. Statement for the record in response to a question from Senator Nunn, 9 February. In: US Congress, Senate Armed Services Committee (1978), part 2, p. 933.

Allgemeine Schweizerische Militärzeitschrift, 1970. Sowjetunion: chemische Waffen. **136:** 595–97.

Allgemeine Schweizerische Militärzeitschrift, 1978. Rüstet die DDR für einen chemischen und biologischen Krieg? 9 September.

Allgemeine Schweizerische Militärzeitschrift, 1979. Gasbomben für die DDR-Armee? No. 7+8, pp. 415–16.

Anderson, J., 1976. UPI wire-story of 10 November from Washington reporting an interview with Lt. Gen. Daniel Q. Graham. See *Washington Post*, Soviet warheads said toxic-tipped, 10 November, for a part of the wire story.

Armed Forces Chemical Journal, 1960. Soviets training with chemical weapons. **14**(6): 25.

Arms Control Today, 1979. Study advises US advance in chemical weapons. **9**(1): 3.

Army Research, Development and Acquisition Magazine, 1978. US adopts UK device: SPAL used for CW defense training. **19**(5): 20, September/October.

Army Research, Development & Acquisition Magazine, 1979 *a*. Army establishes Toxic/Hazardous Materials Agency. **20**(1): 15, January/February.

Army Research, Development & Acquisition Magazine, 1979 *b*. Jet exhaust powers new decontamination system prototype. **20**(3): 11, May–June.

Army Reserve Magazine, 1976. Special report. The silent threat: CB war. September/October, pp. 23–26.

Association of the US Army, 1976. Chemical warfare—a military reality. AUSA position paper, 24 November.

Augustine, N. R. (US Assistant Secretary of the Army for Research & Development), 1975. Statements before the House Armed Services Committee, 7 March. In: US Congress, House Armed Services Committee (1975), part 4, pp. 4190 and 4196.
Augustine, N. R. (US Assistant Secretary of the Army for Research & Development) and Cooksey, *Lt. Gen.* H. H. (Deputy Chief of Staff for Research, Development and Acquisition, US Dept of the Army), 1975. Prepared statement on the FY 1976 and transitional quarter Army R & D budget. In: US Congress, Senate Armed Services Committee (1975), part 4, p. 1781.

Babushkin, *Maj. Gen.* A. (retd), 1978. Improving the Chemical Service during the war. *Voyenno-istoricheskiy Zhurnal*, **1978** (7): 87–95, July (in Russian).
Battista, A. R., 1977. *Defense R & D: Issues and Ideas for the 95th Congress.* A presentation for the House Armed Services Committee by one of its Professional Staff Members, 3 February. In: US Congress, House Armed Services Committee (1977), part 3(1), pp. 5–78, on p. 53.
Bay, *1st Lt.* A., 1978. Defense against chemical attack. *Armor*, **87** (3): 40–45, May–June.
Black, W. E. (US Army Chemical Corps Intelligence Agency), 1961. Statement before the House Defense Appropriations Subcommittee, 23 March. In: US Congress, House Appropriations Committee (1961), part 4, p. 238.
Binder, D., 1978. US re-emphasizing chemical warfare. *New York Times*, 5 June.
Blanton, *Maj. Gen.* C. C. (Director of Budget, US Dept of the Air Force), 1976. Statements for the record in response to questions from Senator McClellan, 4 March. In: US Congress, Senate Appropriations Committee (1976), part 3, pp. 255–56, 400–401, 410–11 and 431–32.
Bonds, R., 1976. Editor, *The Soviet War Machine.* London: Salamander.
Bonds, R., 1978. Editor, *The US War Machine.* London: Salamander.
Brown, *Major* F. J., 1968. *Chemical Warfare: A Study in Restraints.* Princeton: Princeton University Press.
Brown, *General* G. S. (Chairman, US Joint Chiefs of Staff), 1975 *a*. *United States Military Posture for FY 1976*, statement to the Congress, January, pp. 114–20.
Brown, *General* G. S. (Chairman, US Joint Chiefs of Staff), 1975 *b*. Letter of 22 April to Rep. Ottinger. In: US Congress, House Appropriations Committee (1975), part 9, p. 250.
Brown, *General* G. S. (Chairman, US Joint Chiefs of Staff), 1975 *c*. Statement before the House Defense Appropriations Subcommittee, 27 February. In: US Congress, House Appropriations Committee (1975), part 1, pp. 190–91.
Brown, *General* G. S. (Chairman, US Joint Chiefs of Staff), 1976. *United States Military Posture for FY 1977.* Statement to the Congress, 20 January, p. 74.
Brown, *General* G. S. (Chairman, US Joint Chiefs of Staff), 1977. *United States Military Posture for FY 1978.* Statement to the Congress, 20 January, p. 88.
Brown, *General* G. S. (Chairman, US Joint Chiefs of Staff), 1978. *United States Military Posture for FY 1979.* Statement to the Congress, 20 January, pp. 89–91.
Brown, H. (US Secretary of Defense), 1978 *a*. Statement before the House Defense Appropriations Subcommittee, February. In: US Congress, House Appropriations Committee (1978 *a*), part 2, p. 828.
Brown, H. (US Secretary of Defense), 1978 *b*. *Rationalization/standardization within NATO—a report to the US Congress*, 28 January. In: US Congress, Senate Armed Services Committee (1978), part 2, pp. 1586–722.

Brown, H. (US Secretary of Defense), 1979. Prepared statement on the FY 1980 defense budget. In: US Congress, House Appropriations Committee (1979), part 1, pp. 19–360.

Carpenter, W. M. *et al.*, 1977. *Evaluation of Chemical Warfare Policy Alternatives 1980–1990.* A contract study for the US Defense Department. Stanford Research Institute, Strategic Studies Center, February. Three volumes, available through NTIS as documents nos. AD A045333, AD A045344 and AD A045345.

Center for Defense Information, 1978. US ground forces: inappropriate objectives, unacceptable costs. *Defense Monitor*, vol. 7, no. 9 (November).

Chamberlain, J. L., 1977. The origins and evolution of United States chemical warfare policy. In: Carpenter *et al.* (1977), vol. 2, pp. 1–42.

Chemical Corps Journal, 1949. Russian aerial release case. **2** (4): 41.

Conway, P. G., 1972. An analysis of decision making on US chemical and biological warfare policies in 1969. Ph.D. dissertation, Purdue University, pp. 63–64.

Cooksey, *Lt. Gen.* H. H. (Deputy Chief of Staff for Research & Development, US Dept of the Army), 1975 *a*. Statements for the record in response to questions from Senator Thurmond, 12 May. In: US Congress, Senate Appropriations Committee (1975), part 2, pp. 490–91.

Cooksey, *Lt. Gen.* H. H. (Deputy Chief of Staff for Research & Development, US Dept of the Army), 1975 *b*. Statement before the House Armed Services Committee, 7 March. In: US Congress, House Armed Services Committee (1975), part 4, p. 4189.

Cooksey, *Lt. Gen.* H. H. (Deputy Chief of Staff for Research & Development, US Dept of the Army), 1975 *c*. Statements before the House Armed Services Committee, 10 March. In: US Congress, House Armed Services Committee (1975), part 4, pp. 4246–48.

Corddry, C. W., 1975. Soviet said to stockpile nerve gas. *The Sun* (Baltimore), 19 June.

Currie, *Brig. Gen.* J. B. (Deputy Director of Programs, Office of the Deputy Chief of Staff for Programs & Resources, US Dept of the Air Force), 1978. Statement for the record in response to a question from Senator Nunn, 8 March. In: US Congress, Senate Appropriations Committee (1978), part 3, p. 2038.

Currie, M. R. (Director of Defense Research and Engineering), 1975. Letter of 11 August to the Chairman of the House Armed Services Committee of the US Congress with enclosed report on *Department of Defense Programs directed towards an Effective CW/BW Defense Posture* as called for in HASC Report 94–199. The enclosure is printed as Appendix 10 in McCullough and Randall (1976).

Daily Telegraph, 1977. Russia "ready for chemical warfare". 21 January.

Davis, R. M. (Deputy Under Secretary of Defense for Research and Advanced Technology), 1978 *a*. Statement before the Senate Armed Services Committee, 9 March. In: US Congress, Senate Armed Services Committee (1978), part 8, p. 6099.

Davis, R. M. (Deputy Under Secretary of Defense for Research and Advanced Technology), 1978 *b*. Prepared statement on the FY budget for the Science and Technology Program of the Defense Department. In: US Congress, House Armed Services Committee (1978), part 3(2), pp. 1628–46, on pp. 1634–35.

Defense & Foreign Affairs Daily, 1978. USSR: chemical warfare report. No. 177, 19 September.

Dehn, R. *et al.*, 1967. *Handbuch für Unteroffiziere des chemischen Dienstes.* Berlin, GDR: Deutscher Militärverlag. This manual complements the earlier *Handbuch für Soldaten des chemischen Dienstes* (1964).

Deutsche Welle, 1968. Stille Waffen. Broadcast of 26 October.

Dick, C., 1979. The growing Soviet artillery threat. *Journal of the Royal United Services Institute* (London), **124** (2): 66–73, June.

Dodd, N. L., 1976. Chemical warfare and the defence precautions available against them [*sic*]. *Defence Materiel*, **1** (3): 65–68.

Dodd, N. L., 1977. Chemical defense equipment. *Military Review*, **57** (11): 17–21, November.

Donnelly, C., 1976. The Soviet Ground Forces. In: Bonds (1976): 154–75.

Donnelly, C. N., 1977. Fighting in built-up areas: a Soviet view—Part II. *Journal of the Royal United Services Institute*, **122** (3): 63–67, on p. 66.

Drugov, Yu. V., 1959. Editor, *Sanitary Chemical Defence.* Moscow: Medgiz (in Russian).

Economist Foreign Report, 1978. Russia arms its allies for chemical war. 15 November.

Erickson, J., 1971. *Soviet Military Power.* London: Royal United Services Institute.

Erickson, J., 1976. Soviet military capabilities. *Current History*, **71**: 97–136, October.

Erickson, J., 1977. Paper presented at a study conference on Soviet Operations, organized in March 1977 by BDM Corporation, McLean, Virginia. Abstracted in Hauger (1977).

Fairhall, D., 1977. Why Porton dropped its guard. *The Guardian* (London and Manchester), 4 April.

Farr, M., 1979 *a*. West must arm says NATO chief. *Daily Telegraph*, 8 September, p. 6.

Farr, M., 1979 *b*. Chemical arms "could prevent nuclear war". *Daily Telegraph*, 20 September, p. 6.

Feir, *Maj. Gen.* P. R. (Assistant Deputy Chief of Staff for Research, Development & Acquisition, US Dept of the Army), 1976. Statement before the House Defense Appropriations Subcommittee, 1 April. In: US Congress, House Appropriations Committee (1976), part 5, pp. 1085–86 and 1133–35.

Feir, *Maj. Gen.* P. R. (Assistant Deputy Chief of Staff for Research, Development & Acquisition, US Dept of the Army), Greiner, E. (Acting Assistant Secretary of the Army for Installations & Logistics) and Jones, *Maj. Gen.* J. R. (Deputy Chief of Staff for Installations & Logistics, US Marine Corps), 1977. Statements before the Senate Armed Services Committee, 22 March. In: US Congress, Senate Armed Services Committee (1977), part 5, pp. 4020 and 4024.

Finan, J. S., 1974. Soviet interest in and possible tactical use of chemical weapons. *Canadian Defence Quarterly*, **4**(2): 11–15.

Foss, C. F., 1978. US ground forces weapons. In: Bonds (1978): 86–101.

Frank, F. R., 1972. US arms control policy making: the 1972 Biological Weapons Convention case. Ph.D. dissertation, Stanford University, p. 110.

Franke, *Major* S., 1976. *Lehrbuch der Militärchemie*, vol. 1. Berlin: Militärverlag der DDR. 2nd edition.

Franke, *Major* S., Franz, P., Grümmer, G. and Warnke, *Oberstleutnant* W., 1976. *Lehrbuch der Militärchemie*, vol. 2. Berlin: Militärverlag der DDR. 2nd edition.

Frankfurter Allgemeine Zeitung, 1979. Sorge Washingtons über einen sowjetischen "Gaskrieg", 17 May.

Gadsby, N., 1968. Gas war build-up by Russia. Lecture reported in *The Times* (London), 14 November.

Ganas, *Lt. Col.* P., 1969. New developments in chemical and biological warfare. *Forces aeriennes françaises*, **24:** 449–75 (in French).

Gardov, V., 1978. An urgent disarmament task. *Novoye Vremya* (Moscow), No. 44, p. 14, 27 October (In Russian).

Gatland, K. W., 1976. Soviet missiles. In: Bonds (1976): 212–31.

General-Anzeiger (Bonn), 1976. Sowjets halten angeblich Nervengas in DDR bereit. 18 June (reporting statements to Washington journalists by Lt. Gen. D. Q. Graham, former Director of the US Defense Intelligence Agency).

German Federal Government, 1970. *White Paper 1970 on the Security of the Federal Republic of Germany and on the State of the German Federal Armed Forces.* Published by the Federal Minister of Defence on behalf of the German Federal Government. Bonn: Press and Information Office of the German Federal Government. Paragraph 68.

Gibbs, *Lt.* L. D., 1978. CBR: what are we waiting for? *Infantry*, **68**(4): 28–31, July–August.

Gilbert, *Maj. Gen.* W. D. (Deputy Director of Engineering & Services, Office of the Deputy Chief of Staff for Programs & Resources, US Dept of the Air Force), 1978. Statements before the House Military Construction Appropriations Subcommittee, 3 March. In: US Congress, House Appropriations Committee (1978 *b*), part 2, pp. 540–41 and 566–70.

Gowing, M., 1974. *Independence and Deterrence: Britain and Atomic Energy 1945–1952.* London: Macmillan, 2 vols. In vol. 1, p. 94.

Groehler, O., 1978. *Der lautlose Tod.* Berlin/GDR: Verlag der Nation.

Ground Defence International, 1979. The new SS-21 is not yet deployed in East Germany. No. 57, September–October, p. 31.

Gunston, W., 1976. Army weapons. In: Bonds (1976): 176–99.

Haig, *General* A. M., Jr (Supreme Allied Commander, Europe, and Commander-in-Chief, US European Command), 1976. Interview published in *Newsweek*, 9 February.

Haig, *General* A. M., Jr (Supreme Allied Commander, Europe, and Commander-in-Chief, US European Command), 1978. Statement before the Senate Defense Appropriations Subcommittee, 3 March. In: US Congress, Senate Appropriations Committee (1978), part 3, p. 1892.

Hauger, J. S., 1977. The chemical warfare threat. *Military Intelligence*, **3**(3): 2–8.

Haworth, D., 1976. Big Soviet nerve gas build-up worries NATO. *The Observer* (London), 14 November, p. 1.

Haworth, D., 1979. Soviet build up worries NATO. *International Herald Tribune*, 16 May.

Hayhoe, B., 1979. Written answer to a Parliamentary question to the Secretary of State for Defence, 12 June. *Hansard (Commons)*, **968:** *158.*

Henry, *Major* B. C., 1979. It's time to face the reality of chemical warfare. *Marine Corps Gazette*, March, pp. 27–30.

Hoeber, A. and Douglass, J. D., 1978. The neglected threat of chemical warfare. *International Security*, **3**(1): 55–82.

Hoffman, K., 1977. An analysis of Soviet artillery development. *International Defense Review*, **10**(6): 1057–61, December.

Hoffmann, M. R. (US Secretary of the Army) and Kerwin, *General* W. T., Jr (US Army Vice Chief of Staff), 1976. Statements before the House Armed Services Committee, 4 February. In: US Congress, House Armed Services Committee (1976), part 1, pp. 968–70.

Hoffmann, M. R. (US Secretary of the Army) and Rogers, *General* B. W. (US Army Chief of Staff), 1977. Prepared statement on the FY 1978 Army budget. In: US Congress, Senate Appropriations Committee (1977), part 1, pp. 640–768.

Hornsby, N., 1976. NATO call for improved defence capability. *The Times* (London), 2 July.

Hylton, A. R., 1972. History of chemical warfare plant facilities in the United States. Midwest Research Institute, Kansas City, 13 November. Volume 4 of the final report on USACDA contract ST–197.

International Defense Review, 1975. The British Army's NBC shelter. **8**(2): 271.

International Defense Review, 1978 *a*. NAIAD nerve gas detector. **11**(8): 327.

International Defense Review, 1978 *b*. Chemical warheads for E. German Army. **11**(8): 1207.

International Defense Review, 1978 *c*. News in brief. **11**(7): 1013.

International Defense Review, 1979 *a*. Soviet chemical threat a NATO priority. **12**(1): 11.

International Defense Review, 1979 *b*. SS-21 deployed with GSFG. **12**(5): 698.

Jacchia, E., 1978. The specter of chemical war. *International Herald Tribune*, 7–8 October.

Jones, *General* D. C. (Chairman US Joint Chiefs of Staff), 1979. *United States Military Posture for FY 1980:* An overview, with supplement prepared by the Organization of the Joint Chiefs of Staff. In: US Congress, House Appropriations Committee (1979), part 1, pp. 364–454.

Karakchiyev, N. I., 1968. *Military Toxicology and Protection against Weapons of Mass Destruction* (In Russian).

Keith, *Lt. Gen.* D. R. (Deputy Chief of Staff for Research, Development & Acquisition, US Dept of the Army), 1978. Statement before the Senate Armed Services Committee, 7 March. In: US Congress, Senate Armed Services Committee (1978), part 8, pp. 6032–33.

Kjellstrom, *Lt. Gen.* J. A. (Comptroller of the Army), 1976. Statements for the record in response to questions from Senator McClellan, 4 March. In: US Congress, Senate Appropriations Committee (1976), part 3, pp. 127–28 and 286–87.

Kjellstrom, *Lt. Gen.* J. A. (Comptroller of the Army), 1977. Statement for the record in response to a question from Senator McClellan, 3 March. In: US Congress, Senate Appropriations Committee (1977), part 3, pp. 710–11.

Latimer, W. D., 1946. Behavior of gas clouds. In: Pierce, W. C. (ed.), *Military Problems with Aerosols and Nonpersistent Gases*. Washington, D.C.: National Defense Research Committee (volume 1 of the Summary Technical Report of Division 10, NDRC).

Le Hénaff, Y., 1979. Les armes de destruction massive et la politique de défense française. *Protection contre les rayonnements ionisants: revue trimestrielle d'information*. No. 79/80.

Lennon, *Brig. Gen.* L. B., 1978. In: Meselson (1978) *passim.*
Lenorovitz, J. M., 1979. USAF trains against chemical warfare. *Aviation Week & Space Technology*, **111**(4): 61–63, 23 July.
Leonard, J. F., 1977. In: Meselson (1978), pp. 54–55 and 58.
Levin, M. Ye., Malinin, G. A., Mandrazhitskiy, M. N., Sinitsyn, V. P. and Fedorov, V. I., 1960. *Defence against Agents of Mass Destruction.* Moscow (in Russian).
Lohs, Kh., 1974. *Synthetische Gifte.* Berlin: Militärverlag der DDR. 4th edition.
Lohs, Kh. and Martinetz, D., 1978. *Entgiftung: Mittel, Methoden und Probleme.* Berlin: Akademie-Verlag (*Wissenschaftliche Taschenbücher* series, No. 181). This has been published in FR Germany as *Entgiftungsmittel—Entgiftungsmethoden* (Brunswick: Vieweg).

Malooley, *Lt. Col.* R. S., 1974. Gas is not a dirty word in the Soviet Army. *Army*, **24**(9): 21–23.
Manets, *Lt. Gen.* F., 1968. The Chemical Warfare Service and its personnel. *Voennyi Vestnik*, **48**(2): 37–41 (in Russian).
Manets, F. I., Sevost'yanov, P. F., Dudnikov, A. F. and Kondrashov, A. A., 1971. *Protection from Mass Destruction Weapons.* Moscow: Voyenizdat (in Russian).
Marine Corps Gazette, 1979. A report on the first annual chemical defense conference. March issue, pp. 31–38.
Marriott, J., 1977. Chemical warfare. *NATO's Fifteen Nations*, **22**(3): 51–65.
Matoušek, J. *et al.*, 1979. *Extrémně Toxické Nízkomolekulární Syntetické Jedy* [Extremely Toxic Low-Molecular Synthetic Poisons]. Hradec Králové: VLVDÚ JEP.
Mayer, A., 1979. The NBC defence of the Bundeswehr. *Wehrtechnik*, April, pp. 22–28 (in German).
McCullough, J. M. and Randall, B., IV, 1976. Chemical and biological warfare: issues and developments during 1975. US Library of Congress, Congressional Research Service, Science Policy Research Division. Report No. 76–30SP, 5 January.
Menaul, *Air Vice Marshal* S., 1978. Chemical and bacteriological warfare: the threat facing NATO. Mimeographed. Foreign Affairs Research Institute (London), 3/1978.
Meselson, M. S., 1978, Editor, *Chemical Weapons and Chemical Arms Control.* Washington, D.C.: Carnegie Endowment for International Peace.
Metzner, R., 1978. The Bundeswehr NBC Defense Research and Development Institute. *Military Technology and Economics*, **2**(4): 50–51.
Meyer, *Lt. Gen.* E. C. (Deputy Chief of Staff for Operations & Plans, US Dept of the Army), 1978 *a*. Statement before the House Defense Appropriations Subcommittee, 2 March. In: US Congress, House Appropriations Committee (1978 *a*), part 4, p. 72.
Meyer, *Lt. Gen.* E. C. (Deputy Chief of Staff for Operations & Plans, US Dept of the Army), 1978 *b*. Statements for the record in response to questions from Senator Nunn, March. In: US Congress, Senate Appropriations Committee (1978), part 3, pp. 2036–37.
Meyer, *Lt. Gen.* E. C., Lennon, *Brig. Gen.* L. B. and Leonard, *Lt. Col.* J. E., 1978. Defense planning for chemical warfare. In: Meselson (1978), pp. 1–5.
Meyer, *Oberstleutnant* R., 1976. ABC-Abwehr: fester Bestandteil der Ausbildung? *Truppenpraxis*, No. 7, pp. 478–80.

Middleton, D., 1972. Soviet nuclear-war gear said to be unequalled in West. *New York Times*, 2 June.

Middleton, D. 1976. Soviet civil defense a concern for NATO. *New York Times*, 11 October.

Middleton, D., 1978. Soviet-Vietnamese treaty may alter sea strategies. *New York Times*, 8 November.

Military Review, 1979. German Democratic Republic: chemical and biological units. **59**(3): 85.

Miller, E. A. (Assistant Secretary of the Army for Research & Development), 1976. Statement before the Senate Armed Services Committee, 25 February. In: US Congress, Senate Armed Services Committee (1976), part 6, pp. 3199–200.

Miller, E. A. (Assistant Secretary of the Army for Research & Development), 1977. Statement for the record in response to a question from Senator McClellan, 18 March. In: US Congress, Senate Appropriations Committee (1977), part 5, pp. 740–41. *Note*: this statement was, apparently, an unusually detailed side-by-side comparison of different components of US and Soviet CW capabilities; but most of it is deleted from the published version.

Miller, E. A. (Assistant Secretary of the Army for Research & Development) and Cooksey, *Lt. Gen.* H. H. (Deputy Chief of Staff for Research, Development & Acquisition. US Dept of the Army), 1977. Prepared statement on the FY 1978 Army R & D budget. In: US Congress, Senate Armed Services Committee (1977), part 8, p. 5318.

Mills, A. K. and Harris, L. E., 1945. Heeresgasschutzschule I, Celle. Combined Intelligence Objectives Subcommittee report CIOS XXIV-49.

Morton, K. D., 1978. Testing equipment for the German troops. *Military Technology & Economics*, **2**(6): 89–94, November/December.

National Defense, 1979. [Announcement:] Symposium on Army Chemical Defense Program (Fort Belvoir, Virginia, 30 and 31 October 1979). Vol. 63, No. 354, p. 66, May–June.

Nature, 1979. Editorial. Chemical warfare: rearmament or disarmament? Vol. 278, p. 293, 22 March.

Neue Zürcher Zeitung, 1975. Modern Soviet weapons. 28 May (in German).

Newsweek, 1976. Periscope: Russia's gas arsenal, 19 April.

Olenchuk, *Maj. Gen.* P. G. (Director of Materiel Acquisition, Office of the Deputy Chief of Staff for Logistics, US Department of the Army), 1974. Statement for the record in response to a question from Rep. Sikes, 9 May. In: US Congress, House Appropriations Committee (1974), part 7, pp. 277–84.

Olenchuk, *Maj. Gen.* P. G. (Assistant Deputy Chief of Staff for Research, Development & Acquisition, US Dept of the Army), 1975. Statements before the House Defense Appropriations Subcommittee, 24 April. In: US Congress, House Appropriations Committee (1975), part 5, pp. 464–71.

Peart, *Lord* (Lord Privy Seal), 1979. Oral answer to a Parliamentary question. *Hansard (Lords)*, 5 February, col. 443.

Pergent, J., 1970. Services des Poudres. *Forces aeriennes françaises*, **24:** 89–102 (in French).

Perry, W. J. (Under Secretary of Defense for Research and Engineering), 1978. Prepared statement on the FY 1979 Defense Department Program for Research, Development & Acquisition. In: US Congress, Senate Armed Services Committee (1978), part 8, pp. 5592–924.

Phillips, *Lt. Col.* W. A. (Office of the Deputy Chief of Staff for Research, Development & Acquisition, US Dept of the Army), 1976. Statements while showing a film to the House Armed Services Committee, 27 February. In: US Congress, House Armed Services Committee (1976), part 5, pp. 1038–40.

Pozdnyakov, V. V., 1956. The chemical arm. In: Liddell Hart, B. H. (ed.), *The Red Army*. London: Weidenfeld & Nicolson, pp. 384–94.

Prentiss, A. M., 1937. *Chemicals in War*. New York: McGraw Hill.

Pretty, R. T., 1977. Editor, *Jane's Weapon Systems 1977*. London: Macdonald and Jane's Publishers Limited.

Richards, W., 1979. Army slowly dumping delirium bombs. *Washington Post*, 18 July.

Roberts, G. and Freeman, C., 1978. Der Spionagefall Frucht (III): Die Vernehmungen beim Staatssicherheitsdienst. *Der Spiegel*, **32**(26): 134–45, on p. 145.

Robinson, J. P. P., 1975*a*. The United States binary nerve-gas programme: national and international implications. *ISIO Monographs* (University of Sussex: Institute for the Study of Organisation), No. 10.

Robinson, J. P. P., 1975*b*. Controls on CW research and development. Paper presented at the 2nd Pugwash Chemical Warfare Workshop, Stockholm.

Robinson, J. P. P., 1977. Should NATO keep chemical weapons? A framework for considering policy alternatives. *SPRU Occasional Paper series* (University of Sussex: Science Policy Research Unit), No. 4.

Robinson, J. P. P., 1979. Chemical warfare, chemical arms limitation and confidence building: a review of the past year, with proposals. Paper prepared for the 7th Pugwash CW Workshop, Stockholm, 13–17 June.

Robinson, J. P. P. and Kaldor, M. H., 1979. Determinants of weapons succession. University of Sussex, Science Policy Research Unit. Final report to the Ford Foundation on grant no. 765–0503.

Rogers, *General* B. W. (US Army Chief of Staff), 1978. Statement for the record in response to questions from Senator Hart, 1 March. In: US Congress, Senate Armed Services Committee (1978), part 2, pp. 1575–76.

Rogers, *General* B. W. (US Army Chief of Staff), 1979*a*. Statement before the House Defense Appropriations Subcommittee, 22 February. In: US Congress, House Appropriations Committee (1979), part 2, p. 711.

Rogers, *General* B. W. (US Army Chief of Staff), 1979*b*. Statement for the record in response to a question from Rep. Addabbo, 22 February. In: US Congress, House Appropriations Committee (1979), part 2, p. 875.

Rosenblatt, D. H., Dacre, J. C., Shiotsuka, R. N. and Rowlett, C. D., 1977. Problem definition studies on potential environmental pollutants. VIII. Chemistry and toxicology of BZ (3-quinuclidinyl benzilate). US Army Medical Bioengineering Research and Development Laboratory, Fort Detrick. Technical Report no. 7710, August.

Rühle, H., 1977. US chemical warfare policy: Nato perspective. In: Carpenter (1977), volume 2, pp. 83–96.

Rühle, H. 1978. Chemische Waffen und europäische Sicherheit 1980–1990. *Europäische Wehrkunde*, **27**(1): 5–10.

Schlesinger, J. R. (US Secretary of Defense), 1975. Statement before the House Defense Appropriations Subcommittee, 27 February. In: US Congress, House Appropriations Committee (1975), part 1, p. 117.
SIPRI, 1971 *a*. *The Problem of Chemical and Biological Warfare*. Volume 1: *The Rise of CB Weapons*. Stockholm: Almqvist & Wiksell.
SIPRI, 1971 *b*. *The Problem of Chemical and Biological Warfare*. Volume 5: *The Prevention of CBW*. Stockholm: Almqvist & Wiksell.
SIPRI, 1973. *The Problem of Chemical and Biological Warfare*. Volume 2: *CB Weapons Today*. Stockholm: Almqvist & Wiksell.
SIPRI, 1975. Binary nerve-gas weapons. In *Chemical Disarmament: New Weapons for Old*, pp. 21–99. Stockholm: Almqvist & Wiksell.
Soldat und Technik, 1968. Auch Kampfstoff-Rüstung der Sowjets. No. 2, p. 69.
Soldat und Technik, 1970. Chemische Waffen im Warschauer Pakt. No. 9, p. 478.
Soldat und Technik, 1971. Eine Vermehrung der chemischen Waffen in der Sowjetunion. No. 6, p. 344.
Soldat und Technik, 1976 *a*. Eine neue mobile Dekontaminierungsanlage. No. 10, p. 546.
Soldat und Technik, 1976 *b*. Die AC-Abwehrtruppe der Warschau Pakt-Heere. No. 9, p. 479.
Der Spiegel, 1969. Die bakteriologischen und chemischen Waffen der Vereinigten Staaten. 1 December, pp. 154–62.
Der Spiegel, 1979. Gift-Affäre. **33**(39): 19–28, 24 September. See also *Der Spiegel*, **33**(40): 7–8, 24–26 and 30, 1 October.
Stepanov, A. A. and Popov, V. N., 1962. *Chemical Weapons and Principles of Antichemical Defence*. Moscow: Voyenizdat (in Russian).
Stepanskiy, G. A., 1966. *A Short Manual on Toxicology*. Moscow: Meditsina Publishing House (in Russian).
Sterlin, R. N., Yemel'yanov, V. I. and Zimin, V. I., 1971. *Chemical Weapons and Defence against them*. Moscow (in Russian).
Stewart, *Lt. Col.* J. A., 1970. REME in a chemical war. *Journal of the Royal Electrical & Mechanical Engineers*, no. 4, pp. 23–25.
Stöhr, *Oberst* R., Kiesslich-Köcher, *Oberstleutnant* H., Gorges, *Oberstleutnant* H., Martin, *Oberstleutnant* B. and Bäsig, M., 1977. *Chemische Kampfstoffe und Schutz vor chemischen Kampfstoffen*. Berlin: Militärverlag der DDR.
Stroykov, Yu. N., 1970. *Medical Aid for Toxic Agent Victims*. Moscow (in Russian).
Stubbs, *Maj. Gen.* M., 1959. Soviets speed production of germ war weapons. *Register and Defense Times*, 9 May, pp. 24–25.
Stubbs, *Maj. Gen.* M., 1963. CBR and the Army reorganisation. *Armed Forces Chemical Journal*, **17**(3): 5–6.

Templeton, *Colonel* J. L., Jr, 1975. A credible chemical defense: fact or fantasy? US Army War College, Carlisle Barracks, student essay, 20 October. Available through NTIS as document AD A024976.
Tice, J., 1976. Chemical Corps revived: Army reacts to evidence of Soviet activity. *Army Times*, 23 August.
Tolmein, H. C., 1978. Das Geheimnis der DDR—ihre Rüstungsindustrie. *Wehrtechnik*, no. 12, p. 54.
Trudeau, *Lt. Gen.* A. G. (Chief of Research & Development, US Dept of the Army), 1960. Statement before the House Defense Appropriations Subcommittee, 14

March. In: US Congress, House Appropriations Committee (1960), part 6, p. 181.

UK, 1976*a*. Disarmament Conference document CCD/502, 2 July.

UK, 1976*b*. Disarmament Conference document CCD/512, 6 August.

UK, 1979. Disarmament Conference document CD/15, 24 April.

US Arms Control & Disarmament Agency, 1979. Chemical warfare. In *Fiscal Year 1980 Arms Control Impact Statements*. Statements submitted to the Congress pursuant to Section 36 of the Arms Control and Disarmament Act, 13 February.

US Army Armament Research and Development Command, 1977. *Laboratory Posture Report FY77*. P–RCS–DRC LDC–101, December.

US Army Edgewood Arsenal, 1975. In: US Senate, Committees on the Judiciary and on Labor and Public Welfare. *Biomedical and Behavioral Research, 1975*. Joint hearings. Washington, D.C.: US Government Printing Office, 1976, pp. 768–73.

US Comptroller General, 1977. Report to the Congress. *Stockpile of lethal chemical munitions and agents: better management needed*, LCD–77–205, 14 September. (This appears to be an unclassified digest of the secret report *US Lethal Chemical Munitions Policy: Issues Facing the Congress*, PSAD–77–84, 21 September.)

US Congress, House Appropriations Committee, 1960. Subcommittee hearings on *Department of Defense Appropriations for 1961*. Washington, D.C.: US Government Printing Office.

US Congress, House Appropriations Committee, 1961. Subcommittee hearings on *Department of Defense Appropriations for 1962*. Washington, D.C.: US Government Printing Office.

US Congress, House Appropriations Committee, 1969. Subcommittee hearings on *Department of Defense Appropriations for 1970*. Washington, D.C.: US Government Printing Office.

US Congress, House Appropriations Committee, 1974. Subcommittee hearings on *Department of Defense Appropriations for 1975*. Washington, D.C.: US Government Printing Office.

US Congress, House Appropriations Committee, 1975. Subcommittee hearings on *Department of Defense Appropriations for 1976*. Washington, D.C.: US Government Printing Office.

US Congress, House Appropriations Committee, 1976. Subcommittee hearings on *Department of Defense Appropriations for 1977*. Washington, D.C.: US Government Printing Office.

US Congress, House Appropriations Committee, 1978*a*. Subcommittee hearings on *Department of Defense Appropriations for 1979*. Washington, D.C.: US Government Printing Office.

US Congress, House Appropriations Committee, 1978*b*. Subcommittee hearings on *Military Construction Appropriations for 1979*. Washington, D.C.: US Government Printing Office.

US Congress, House Appropriations Committee, 1979. Subcommittee hearings on *Department of Defense Appropriations for 1980*. Washington, D.C.: US Government Printing Office.

US Congress, House Armed Services Committee, 1975. *Hearings on Military Posture and H.R.3689 [H.R.6674] Department of Defense Authorization for Appropriations for Fiscal Year 1976*. HASC No. 94–8. Washington, D.C.: US Government Printing Office.

US Congress, House Armed Services Committee, 1976. *Hearings on Military Posture and H.R.11500 [H.R.12438] Department of Defense Authorization for Appropriations for Fiscal Year 1977*. Washington, D.C.: US Government Printing Office.
US Congress, House Armed Services Committee, 1977. *Hearings on Military Posture and H.R.5068 [H.R.5970] Department of Defense Authorization for Appropriations for Fiscal Year 1978*. Washington, D.C.: US Government Printing Office.
US Congress, House Armed Services Committee, 1978. *Hearings on Military Posture and H.R.10929 Department of Defense Authorization for Appropriations for Fiscal Year 1979*. Washington, D.C.: US Government Printing Office.
US Congress, Senate Appropriations Committee, 1975. Subcommittee hearings on *Department of Defense Appropriations for Fiscal Year 1976*. Washington, D.C.: US Government Printing Office.
US Congress, Senate Appropriations Committee, 1976. Subcommittee hearings on *Department of Defense Appropriations for Fiscal Year 1977*. Washington, D.C.: US Government Printing Office.
US Congress, Senate Appropriations Committee, 1977. Subcommittee hearings on *Department of Defense Appropriations for Fiscal Year 1978*. Washington, D.C.: US Government Printing Office.
US Congress, Senate Appropriations Committee, 1978. Subcommittee hearings on *Department of Defense Appropriations for Fiscal Year 1979*. Washington, D.C.: US Government Printing Office.
US Congress, Senate Armed Services Committee, 1975. Hearings on *Fiscal Year 1976 and July-September 1976 Transition Period Authorization for Military Procurement, Research and Development, and Active Duty, Selected Reserve, and Civilian Personnel Strengths*. Washington, D.C.: US Government Printing Office.
US Congress, Senate Armed Services Committee, 1976. Hearings on *Fiscal Year 1977 Authorization for Military Procurement, Research and Development, and Active Duty, Selected Reserve, and Civilian Personnel Strengths*. Washington, D.C.: US Government Printing Office.
US Congress, Senate Armed Services Committee, 1977. Hearings on *Fiscal Year 1978 Authorization for Military Procurement, Research and Development, and Active Duty, Selected Reserve, and Civilian Personnel Strengths*. Washington, D.C.: US Government Printing Office.
US Congress, Senate Armed Services Committee, 1978. Hearings on *Department of Defense Authorization for Appropriations for Fiscal Year 1979*. Washington, D.C.: US Government Printing Office.
US Dept of the Army, 1971*a*. *Chemical Reference Handbook*. Field manual FM 3–8, January 1967, plus Changes nos. 1–4 through to 18 October 1971.
US Dept of the Army, 1975. *Handbook on Soviet Ground Forces*. FM 30–40, 30 June.
US Dept of the Army, 1976*a*. *Operations*. Field manual FM 100–5, 1 July.
US Dept of the Army, 1976*b*. Justification Book for the FY 1977 Army RDT & E estimates, p. 41. In: US Congress, House Appropriations Committee (1976), part 3, p. 238.
US Dept of the Army, 1977*a*. *US Army Activity in the US Biological Warfare Program*. 24 February, Vol. 1, pp. 4.2 and 5.2.
US Dept of the Army, 1977*b*. Chemical warfare. *Commanders Call*, January–February (DA–Pam–360–831): 3–13, on p. 8.

US Dept of the Army, 1977 *c*. *NBC Defense*. Field manual FM 21–40, 14 October.
US Dept of the Army, 1977 *d*. *Fire Support in Combined Arms Operations*. Field manual FM 6–20, 30 September.
US Dept of Defense, 1969. Statement for the record in response to a question from Rep. Sikes, 9 June. In: US Congress, House Appropriations Committee (1969), part 6, pp. 136–37.
US Dept of Defense, 1976 *a*. Statement for the record in response to a question from Senator Thurmond to the Chairman of the Joint Chiefs of Staff, 29 January. In: US Congress, Senate Armed Services Committee (1976), part 1, p. 558.
US Dept of Defense, 1976 *b*. Statement for the record in response to a question from Senator McClellan, 2 February. In: US Congress, Senate Appropriations Committee (1976), part 1, p. 510.
US Dept of Defense, 1977. Statement for the record in response to a question from Senator Bartlett, 8 March. In: US Congress Senate Armed Services Committee (1977), part 3, pp. 2221–24.
US Dept of Defense, 1978. Testimony before the House Military Construction Appropriations Subcommittee during hearings on "Security of Nuclear and Chemical Weapons Storage", 23–24 February. In: US Congress, House Appropriations Committee (1978 *b*), part 2, pp. 137–336.
US Joint Chiefs of Staff, Joint Intelligence Group, 1949. Memorandum for the Joint Strategic Plans Group: Estimate of Soviet capabilities for employing biological and chemical weapons. JIGM–80, in enclosure of JIG 297/3, 27 January.
US National Security Council, 1970. United States policy on toxins. National Security Decision Memorandum no. 44, 20 February.
US Strategic Bombing Survey, 1945. Oil, Chemicals and Rubber Division, Ministerial Report no. 1. *Powder, Explosives, Special Rocket and Jet Propellants, War Gases and Smoke Acid*. November 1945.
Velenets, I. S., Drouzhinin, L. M. and Pavlov, Ye. I., 1968. [The Protection of Subunits from Weapons of Mass Destruction]. Moscow: Voenizdat (in Russian).
Volz, A., 1976. Soviet chemical warfare capability. *Radio Liberty Research* (Munich) RL 437/76, 11 October.
Watkins, T. F., 1968. Chemical warfare. In: Watkins, T. F., Cackett, J. C. and Hall, R. G. (eds), *Chemical Warfare, Pyrotechnics and the Fireworks Industry*. Oxford: Pergamon.
Watson, *Colonel* G. G. (Chief of Chemical and NBC Defense Division, Directorate of Strategy & Plans, Office of the Deputy Chief of Staff for Operations, US Dept of the Army), 1978. Briefing on chemical warfare for the House Armed Services Committee (1978), part 3(1), pp. 694–707.
Wehrtechnik, 1977. Schutzbekleidung gegen chemische Kampfstoffe. No. 5, p. 98.
Weinraub, B., 1977. NATO fears Soviet gas warfare. *New York Times*, 17 May 1977.
Weinstein, A., 1978. Kein Wort über chemischen Krieg. *Frankfurter Allgemeine Zeitung*, 24 October.
Weyand, *General* F. C. (Army Chief of Staff), 1975. Statement before the House Armed Services Committee, 24 February. In: US Congress, House Armed Services Committee (1975), part 1, pp. 536–37.
Wood, D. and Pengelley, R. B., 1977. Nuclear defence in the European Environment. Part 1: The threat. *International Defense Review*, **10**(4): 631–34.
Zhuk, N. M. and Stroykov, Y. N., 1972. *Protecting the Population against Chemical Weapons*. Moscow (in Russian).

Destruction of US chemical weapons production and filling facilities

R. MIKULAK[1]

US Arms Control and Disarmament Agency, Washington, D.C., USA

Abstract. Complete and effective prohibition of chemical weapons requires not only destruction of existing stockpiles, but also elimination of facilities for producing chemical warfare agents and for filling chemical munitions. To date, however, there has been little, if any, technical discussion of the task of destroying or dismantling such facilities.

In order to provide necessary background, this paper begins with a description of the steps in production of the nerve agent sarin and of the US facilities involved. The procedures for destruction and dismantling of such facilities are then outlined and implications for verification considered.

I. Introduction

This paper is intended to provide a general and preliminary discussion of the task of destroying or dismantling facilities designed or used for production of the means of chemical warfare. For illustrative purposes, the discussion deals with a representative nerve agent, sarin, and with US facilities and experience. In view of the differences in production processes and in safety and environmental regulations among countries, the procedures required may vary somewhat for different agents and from country to country. There is no apparent reason, however, why this factor should affect the generality of the conclusions drawn from US experience.

II. Synthesis of sarin (agent GB)

The process steps required for the production of methylphosphonic dichloride (dichlor) by the dimethyl hydrogen phosphite (DMHP) method, and the follow-on conversion of dichlor to sarin, are discussed below.

The dimethyl hydrogen phosphite (DMHP) process—steps 0 to III and $POCl_3$ reduction

Briefly, the reaction steps required for the production of the intermediate methylphosphonic dichloride (dichlor) are as follows:

[1] The author wishes to acknowledge very helpful discussions with personnel of the US Army Chemical Systems Laboratory and the US Army Toxic and Hazardous Materials Agency, both located at Aberdeen Proving Ground, Maryland. The opinions expressed in this paper are those of the author and do not necessarily represent the views of any agency of the US Government.

1. In the initial step (step 0), phosphorus trichloride (PCl_3) is made by reacting chlorine with elemental phosphorus according to the following reaction:

$$2P + 3Cl_2 \rightarrow 2PCl_3$$

2. Phosphorus trichloride is reacted (step I) with methanol (CH_3OH) to produce dimethyl hydrogen phosphite, $(CH_3O)_2POH$, as follows:

$$PCl_3 + 3CH_3OH \rightarrow (CH_3O)_2POH + 2HCl + CH_3Cl$$

3. The dimethyl hydrogen phosphite is pyrolysed (step II) to produce a pyro mix containing methyl hydrogen methylphosphonate, *bis*-hydroxymethylphosphonyl oxide, and derivatives of pyrophosphoric acid (alkylation). The basic reaction is:

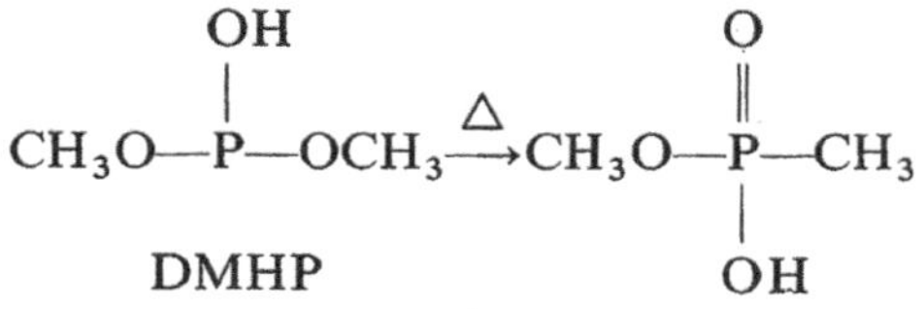

and

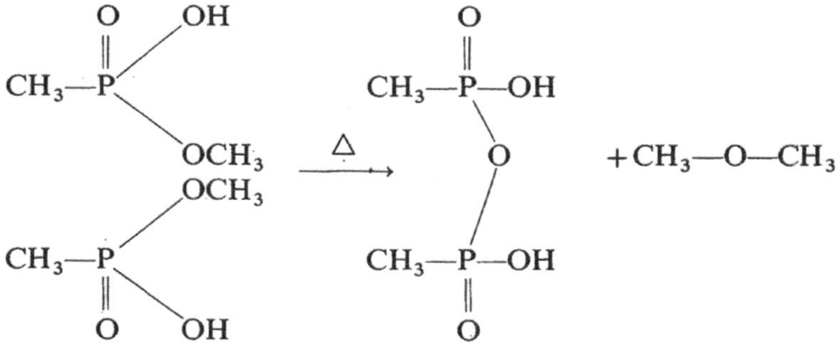

4. The pyro mix is chlorinated (step III) using chlorine and phosphorus trichloride to form methylphosphonic dichloride (dichlor):

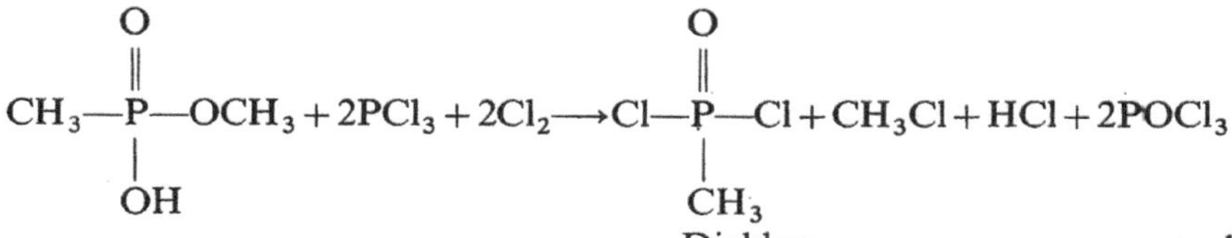

and

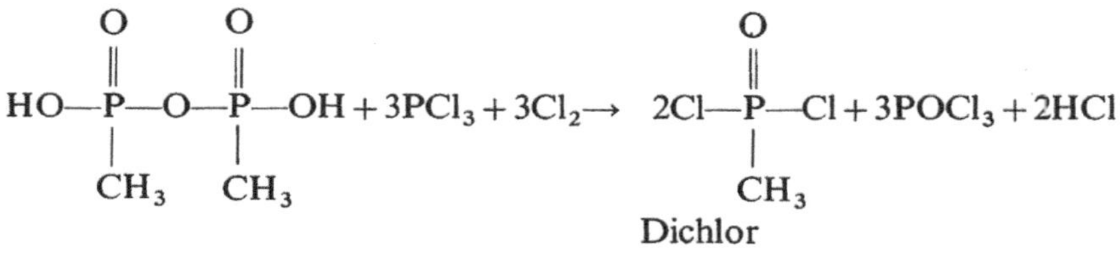

A further step is required to reduce by-product phosphorus oxychloride to phosphorus trichloride, which is subsequently reused. This is accomplished by passing the $POCl_3$ through an incandescent bed of charcoal. It proceeds according to the following reactions:

$$POCl_3 + C \longrightarrow PCl_3 + CO;$$

$$2POCl_3 + C \longrightarrow 2PCl_3 + CO_2$$

Production of sarin from intermediate

The steps required for conversion of the intermediate methylphosphonic dichloride to sarin are as follows:

1. Methylphosphonic dichloride is reacted with liquid hydrofluoric acid to produce an approximately equimolar mixture of methylphosphonic dichloride and methylphosphonic difluoride according to the following reaction:

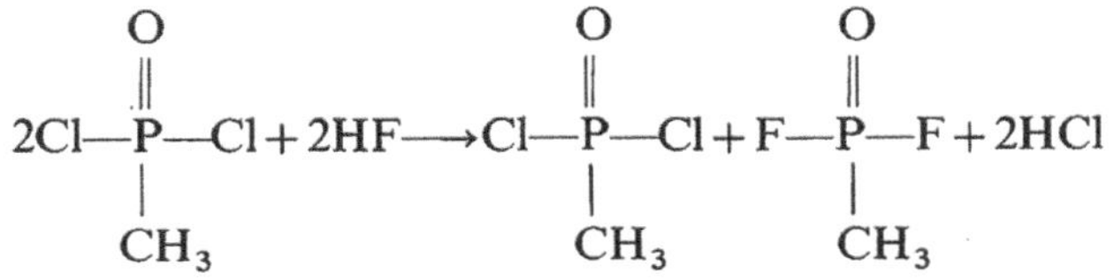

2. The liquid mixture from the first reaction is reacted with isopropanol to form agent as follows:

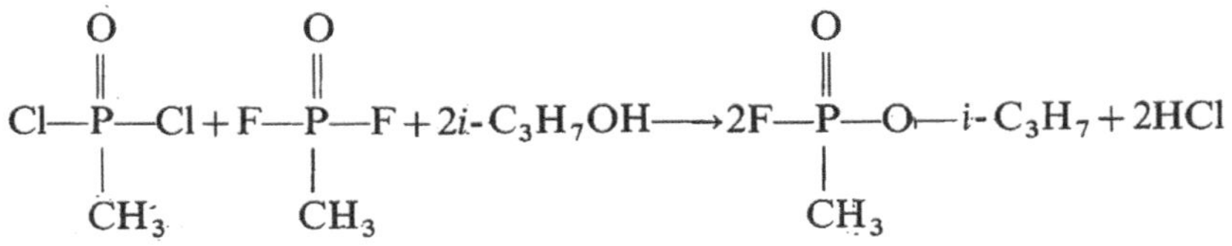

III. Chemical weapons production and filling facilities

Phosphate Development Works

In the United States, methylphosphonic dichloride (dichlor), the intermediate in sarin production as described above, was once synthesized at the Phosphate Development Works (PDW). This plant, which has been inactive since 1964, is located on 18 hectares of land on the property of the Tennessee Valley Authority at Muscle Shoals, Alabama.

Table 1 lists the key items of equipment and utilities located on the PDW site.

Table 1. Phosphate Development Works—key equipment and utilities

Step 0	**Utilities** (*continued*)
(8 reactors)	Water cooling tower
Manufacturing (PCl_3)	Water stand pipe storage
Control house	Steam distribution
Phosphorus storage	Water supply and distribution
PCl_3 storage	Electric distribution
	Compressed air distribution
Step 1	Instrument air distribution
(2 trains)	Sanitary and storm sewage
Manufacturing (DMHP)	
Control house	**Items common to several steps**
Compressor house	HCl storage
Methyl alcohol storage	H_2SO_4 storage
Methyl chloride storage	Caustic storage, 18%
DMHP storage	Chlorine evaporators
Phenylcyclohexane storage[a]	Chlorine storage
	Caustic storage, 50%
Step 2	Inert gas holder
(4 units)	HCl neutralization
Manufacturing building (pyro)	
Control house	**Auxiliary facilities**
CCl_4 storage[a]	Warehouse
Pyro storage	Utility building
	Gate house
Step 3	Diesel generator building
(8 reactors)	Overhead pipe trestle
Manufacturing building (dichlor)	Administration building and laboratory
Control house	Fencing
Refrigeration building	Roads, walks, parking areas, culverts, etc.
Dichlor storage	Process distribution
$POCl_3$ storage	Railroads
PCl_3 storage	
High temperature methanation equipment	**By-product $POCl_3$ reduction**[b]
Control house	Reduction facility
Gas preparation	Control house
Utilities	
Heat medium and inert gas	**Semi-permanent buildings**
Water pumping section	Administration building
Chemical sewage disposal	Instrument laboratory
	Equipment storage
	Paint storage
	Warehouse

Notes:
[a] No longer used in process.
[b] These facilities have been removed except for three transformer buildings.

Rocky Mountain Arsenal

Dichlor produced at PDW was converted to sarin at Rocky Mountain Arsenal, which is located approximately 16 kilometres north-east of Denver, Colorado. Since the arsenal was constructed, adjacent areas have been developed, so that now the 70-square-kilometre Arsenal Grounds are virtually

Figure 1. Rocky Mountain Arsenal—sarin production and filling area

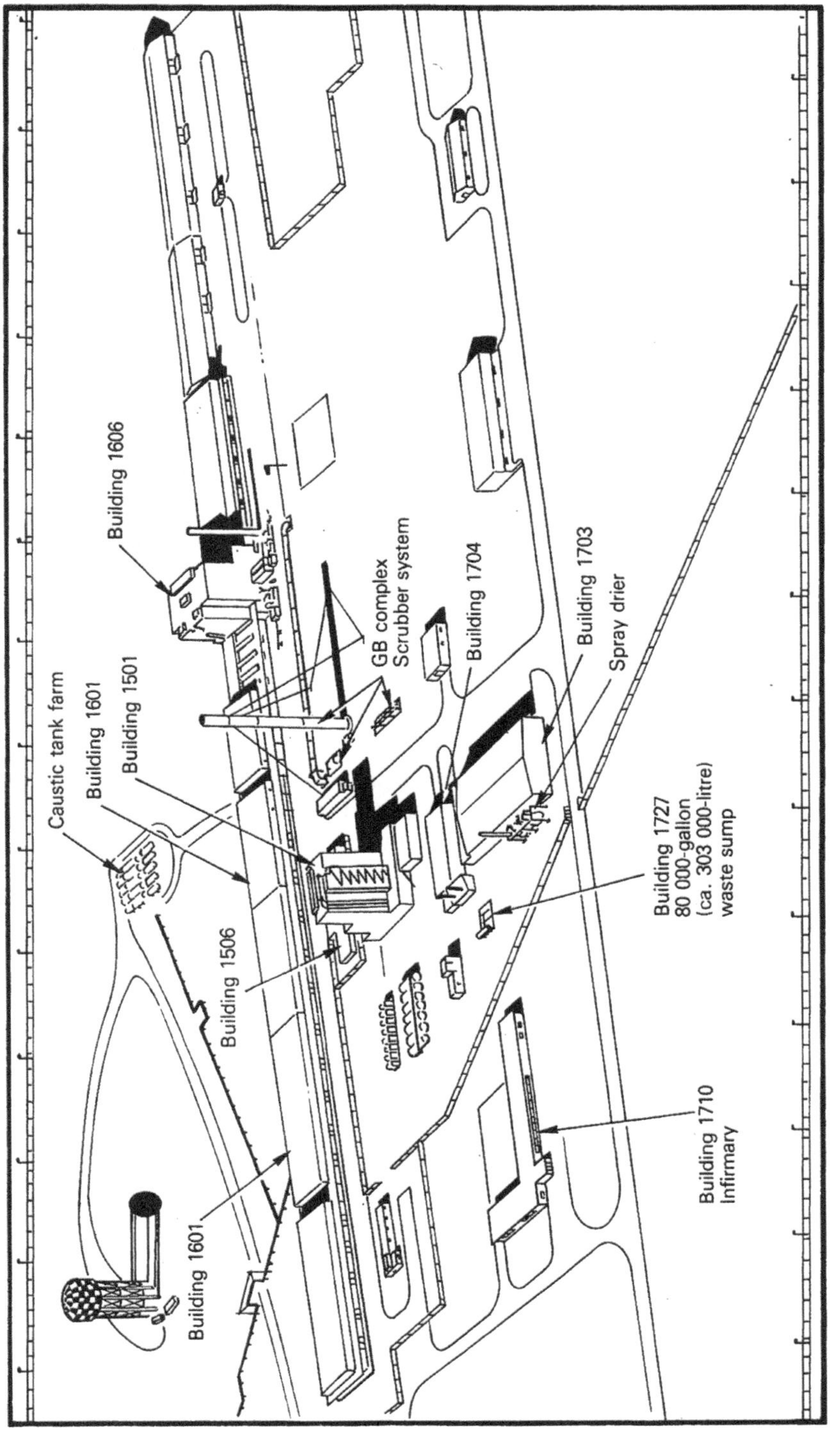

surrounded by populated areas. Stapleton International Airport is located immediately to the south and, in fact, has annexed some land inside the original south boundary of the arsenal.

The sarin production and filling area occupies approximately 125 hectares in the north-central section of the arsenal. Construction of the sarin production facilities began in 1951 and was completed in 1953. The chemical production facilities were placed in "layaway" or stand-by status in the period 1955–57, and the filling facilities in 1968.

The principal buildings associated with sarin production and filling are shown in figure 1 and are discussed below.

Agent production (*Building 1501*)

The building in which sarin was produced from dichlor is 37 metres (six storeys) high, 70 metres long and 48 metres wide. It is windowless and is constructed of reinforced concrete. The basic structure is divided vertically into three similarly equipped operating bays, each originally designed to operate as a complete unit for sarin production. During the mid-1970s, one bay was converted for sarin demilitarization operations.

Agent storage (*Building 1506*)

This building was used for storage of newly produced sarin before the agent was filled into munitions or one-ton containers. It consists of a small, single-storey, concrete block structure above ground and an underground structure housing ten 10 000-gallon (*ca.* 37 900-litre) storage tanks and the associated pumping and transfer systems. Each tank is 8 metres long and 2.5 metres in diameter.

Filling (*Building 1606*)

This building was originally used for filling and packing sarin-filled munitions containing explosive. It is 183 metres long, approximately 30 metres wide and, except for the central portion, one storey high. The ends of the building are of cinder-block construction and house equipment rooms for filling and processing sarin munitions without explosives. Explosive cubicles are located across the centre portion of the building. Two sets of six cubicles exist, each enclosed by steel-reinforced concrete walls, 0.6-metre thick. Building 1606 was used in the mid-1970s for use in chemical munition demilitarization operations. Much of the original filling equipment was removed at that time.

Filling (*Buildings 1601 and 1601A*)

These buildings, which abut Building 1606, are one-storey steel-frame structures with cinder-block walls. Together, they are approximately 37 metres wide and more than 183 metres long. One section houses pumps and control equipment used in transferring chemicals to and from the underground tanks in Building 1506. Another section was used for filling artillery projectiles.

IV. Procedures for destruction or dismantling

The purpose of requiring production and filling facilities to be destroyed or dismantled would be to ensure that CW agent production or filling could not be resumed. For this purpose it would be important to ensure that the components could not be reassembled.

This does not mean that everything at the site would have to be destroyed. Buildings housing administrative and support services (for example, offices, laboratories, warehouses and utilities) could be allowed to remain. Some pieces of production equipment might also be salvaged for other uses.

The procedures for destroying or dismantling a particular production stage would be governed to a considerable extent by the toxicity of the chemicals involved at that stage. Extreme precautions would be required for tearing down the stage where a supertoxic chemical was produced. However, less stringent precautions would be needed for tearing down stages which produced precursors to supertoxic chemicals, since the chemical residues present would be much less toxic than the supertoxic chemical itself.

In destroying or dismantling a facility it might make sense to begin with the supertoxic areas. In this way the areas of greatest potential hazard to workers, which are also perhaps the most important areas for agent production, would be eliminated first. Furthermore, dismantling of the supertoxic areas is likely to proceed more slowly than dismantling of other areas. From a practical standpoint, it would be prudent to begin the slowest process first.

Supertoxic chemical facilities

The first step in actual dismantling/destruction operations would be to flush a decontaminant (for example, 20% aqueous sodium hydroxide solution in the case of sarin) through the process equipment. If the facility had been closed down sometime before the beginning of the destruction operation, the equipment would have been flushed at that time. Nonetheless, it would be essential to repeat the process before attempting to remove the equipment. At about the same time, efforts would be made to remove surface contamination from process equipment and from the walls, ceilings and floors of the area where the equipment is located. The surface decontamination effort would be slow and very labour-intensive and could only be accomplished by workers outfitted in cumbersome protective clothing. These suits would seriously impede movement and would impose such a heat burden that they could only be worn for short periods.

Once all gross surface contamination had been eliminated, the actual destruction/dismantling work would begin. To the extent possible, equipment would be taken apart by removing bolts and other fasteners. However, much of the equipment might have to be cut apart with acetylene torches. Large reactors and tanks would have to be cut into small sections.

All personnel involved in the dismantling/destruction operation would have to wear protective masks and clothing. Past experience has shown that agent will probably have seeped into areas (for example, gaskets) which surface decontamination efforts do not reach. Furthermore, small quantities of agent trapped in metal surfaces would be vaporized by cutting torches, resulting in a toxic vapour hazard.

As metal pieces were removed they would be placed in a decontaminant bath for a number of hours, rinsed, and then heated to approximately 540°C in a metal parts furnace. Items so treated would be free of toxic chemicals and could be sold as metal scrap. A similar treatment process has already been used in the United States for demilitarization of certain chemical munitions.

Non-supertoxic chemical facilities

In general, the steps in destruction/dismantling of non-supertoxic facilities would parallel those for supertoxic facilities. Equipment would be flushed with a decontaminant solution and surfaces would be scrubbed down. Then the equipment would be taken (or cut) apart and further decontaminated. However, heating in a furnace would probably not be required.

While the overall procedures would be similar in the two cases, extraordinary precautions (for example, elaborate protective clothing) would not necessarily be required at non-supertoxic facilities. Standard industrial practices and precautions would be sufficient. As a result, the destruction/dismantling process would be carried out more rapidly and at lower cost.

Demolition of production and filling buildings

Process equipment for low-toxicity operations with non-supertoxic precursor chemicals is likely to be located in the open or in buildings of light construction. Whatever demolition was necessary could be readily carried out using standard demolition equipment and procedures.

Supertoxic chemicals of low vapour pressure (for example, VX) are likely to have been produced and filled into munitions in buildings of light construction. However, the relatively volatile, supertoxic chemical sarin was produced by the USA in a massive reinforced concrete structure, which will be considerably more difficult to demolish. In the USA, sarin was filled into explosive-containing munitions in a massive reinforced concrete structure, while munitions not containing explosive were filled in a building of light construction.

To destroy any residual toxic chemicals, rubble from demolition could first be treated in a high-temperature rotary kiln of the type used in the cement industry and then buried in an ordinary landfill. Where environmental regulations and the toxicity of the chemicals involved permit, the rubble could be taken directly to a special landfill of the type used for disposing of hazardous industrial chemical wastes. As an indication of the amount of material to be disposed of, it has been estimated that demolition of the sarin facilities at Rocky Mountain Arsenal would produce 114 000 tonnes of rubble.

Time and manpower requirements

Since it is not possible to make any accurate estimate of costs without specific facilities in mind, only a very general estimate of time and manpower requirements and costs for destruction/dismantling can be made. Generally speaking, it appears that many hundreds of man-years of effort will be required to eliminate a large-scale production facility.

The scale of effort required can be illustrated by US experience in dismantling equipment at Rocky Mountain Arsenal once used for demilitarizing the M34 Agent GB cluster bomb. This effort involved 350 men working for 2–3 months (50–60 man-years) and cost roughly $2 million. As a very crude estimate, destruction/dismantling of the remaining chemical weapons production and filling facilities at Rocky Mountain Arsenal would require an effort at least an order of magnitude greater. At least two years would be required to accomplish such an effort, not including time needed for planning and satisfying environmental regulations.

Because of the hazard involved it is likely that US supertoxic facilities would be torn down by specially trained government personnel. The non-supertoxic facilities, on the other hand, could probably be torn down by a commercial contractor experienced in such work.

V. Monitoring of destruction

As a general rule, activities which occurred outdoors could be monitored directly and from a distance. Remote monitoring of activities taking place within a building would be considerably more difficult since it could be accomplished only indirectly, if at all.

The first task which would have to be faced when the facilities were initially declared would be to establish that they were, in fact, chemical weapons production and filling facilities. The simplest and most reliable way to accomplish this would be through an on-site visit by technical experts. If the nature of the declared facility had been confirmed, conclusions from subsequent monitoring activities could be drawn with much greater confidence.

In the initial stages of actual destruction operations some or all of the following activities might be observed:

(*a*) delivery or storage of large quantities of chemicals used in decontamination;

(*b*) disposal in open ponds of liquid wastes resulting from flushing of process equipment;

(*c*) installation and operation of equipment for spray-drying of liquid wastes;

(*d*) installation and operation of a metal parts furnace;

(*e*) accumulation of piles of metal scrap.

If much of the process equipment were located in the open, as at PDW, the destruction/dismantling could be observed directly. However, for facilities, such as Rocky Mountain Arsenal, where the process equipment is housed in buildings, most of the destruction/dismantling could only be monitored indirectly, such as through observing the accumulation of piles of scrap metal and equipment. If scrap piles were observed remotely, they could be compared with the equipment noted on previous on-site visits. This comparison would help to provide confidence that the process equipment had been destroyed rather than carted away intact. Even at a closed-in facility, some dismantling activities might be observed directly. For example, at Rocky Mountain Arsenal, removal of external storage tanks could be easily seen.

Demolition of buildings, which would be relatively easy to monitor from a distance, would, of course, provide the simplest and most conclusive evidence that the production and filling facility had been eliminated.

Finally, it is possible that remote monitoring might be facilitated by prior agreement on procedures to be employed in destruction and dismantling. Provisions for agreed dismantling procedures have already been incorporated in Articles VIII and XIII of the 1972 US–Soviet treaty on anti-ballistic missile systems.

Destruction or conversion of chemical warfare agents: possibilities and alternatives

Kh. LOHS
Academy of Sciences of the GDR, Research Department of Chemical Toxicology, Leipzig, GDR

Abstract. The production and stockpiling of chemical warfare (CW) agents, intermediates and binary components continues as a part of the arms race. Fundamental approaches to the problem of destruction of CW agents and related compounds, and to the verification of such destruction, are still widely divergent. Destruction methods are technically feasible for virtually all militarily significant CW agents. The difficulties encountered in the destruction operations relate largely to the quantities to be disposed of and to the problem of verification.

Up to now there has been no collaboration between chemists from the military and civilian sectors on the problem of conversion of CW agents. Although conversion is a limited alternative to destruction, it is nevertheless a worthwhile endeavour. Most of the base products and intermediates in CW-agent manufacture can find application in the civilian chemical industry—for example, in the production of pesticides, pharmaceuticals, dyestuffs and plastics. The conversion of CW agents or their intermediate products can also lead to the development of new synthetic components, which have hitherto been only of academic interest or have not at all been described in the literature. The conversion of CW agents and of their base materials and intermediate products can be a technically feasible alternative to the unconditional destruction of these compounds.

I. The present situation

The arms race continues. A variety of chemical weapons are being produced and stockpiled in large quantities in the arsenals of the weapons of mass destruction. Among these chemical weapons are the so-called binary weapons, developed in recent years by the United States and its allies.

For the past 10 years or so, chemical disarmament has been a subject of intense debate, first at the Conference of the Committee on Disarmament (CCD) in Geneva and then at the negotiating body which replaced it in 1979—the Committee on Disarmament (CD). Despite these efforts, no international agreement has yet been reached. Indeed, since the conclusion of the convention on the prohibition of the development, production and stockpiling of biological and toxin weapons and on their destruction that was opened for signature in April 1972, there has been very little progress towards the conclusion of a similar treaty on chemical weapons. The fundamental approaches of the Eastern and Western blocs to the problems of CW agents are still widely divergent and unbridgeable (Nerlich, 1977; World Federation of Scientific Workers, 1977; Lohs, 1978; Meissner and Lohs, 1978; Meselson, 1978).

II. Possibilities for the destruction of CW agents

The goal of the destruction of CW agents is to make these compounds unusable for military purposes and to reduce their high toxicity so that they are no longer a hazard to man and the environment. Thus, the detoxification of CW agents is tantamount to their destruction. Of course, the possibility remains that the cleavage products of destruction may again be used for the manufacture of CW agents. This possibility is, however, only of academic interest, since it would be far less costly and technically less cumbersome to manufacture the CW agents anew from fresh starting materials.

It is not necessary to enlarge here on the various methods for the destruction of CW agents, since comprehensive descriptions are available in the literature (e.g., SIPRI, 1975, 1978, 1979).

Destruction methods are available for virtually all militarily significant CW agents. Special difficulties may arise only if the CW agents are already loaded into munitions, particularly when the latter have become defective. The danger increases if the munitions are already provided with a fuse, or detonator. However, even under such circumstances, specific dismantling and destruction methods are available. The difficulties encountered in the destruction operations lie mainly in the quantities of CW agents to be eliminated and in the problem of verification of these operations.

Former methods for disposing of old stockpiles of CW agents, such as dumping in the sea or in deep lakes or burying underground—as was done after World Wars I and II—can no longer be used in view of the environmental damage caused.

Those who continue to amass chemical munitions contrary to all military reason must realize that the destruction costs will by far outstrip the production costs. However, even the most expensive destruction operation is less costly than involvement in foolhardy military ventures based on the use of CW agents.

It is a plain fact that the German Democratic Republic has not produced CW agents at any time. Our experience in the field of destruction of CW agents is the result of the work we have done in the GDR after World War II (mainly between 1955 and 1961) in destroying all the CW-agent plants of Hitler's *Wehrmacht* and in detoxifying the CW agents filled in shells and kept in bulk stocks and also dispersed in the environment.

New technical solutions are likely to be necessary in the future for detoxification under special conditions. Further research and development will be necessary. The paper on the problem of catalytic detoxification presented by the GDR at the Conference of the Committee on Disarmament in 1976 offers some helpful hints (GDR, 1976: 3):

> *Metal-catalyzed decomposition*
> The catalytic splitting of organophosphorus esters especially by heavy metals has been known for some 20 years. The first observations were made in connexion with the loss of effect of phosphoric ester preparations in cupriferous containers. Apart from [this], biochemical studies have shown that organophosphorus esters are catalytically splittable by a number of metals, e.g. even by lanthanides. Especially the hydroxy-aquo complex as well as the amminohydroxo and alkylamino complexes of copper have shown to be extraordinarily effective for splitting acyl-substituted phosphoric and phosphonic esters.
>
> In this field the highly effective tetraalkyldiamino-copper complexes are practically important.
>
> [This] is suggested by the rapidity and completeness of catalytic ester splitting achieved by these copper complexes, because the solubility of such complexes and the possibility of fixing these compounds to carriers should promote investigations of detoxification for concentrated CW agents of the G- and V-type.

Catalytic splitting may serve as an example of a new and more effective method for destroying such compounds as sarin and soman. New synthetic enzymes have been discovered in the past 4–5 years. It may be useful to test these catalysts in work on the destruction of the highly toxic organophosphorus CW agents.

It is not intended here—nor would it be appropriate—to make any recommendations as to which of the above-mentioned detoxification methods for CW agents would be specially suitable from the technical or economic points of view.

III. Conversion of CW agents as an alternative to destruction

In view of the aggregate cost of raw materials, development and production as well as of adequate safety provisions during manufacture and storage, it is no wonder that CW agents are highly expensive products. CW agents and their intermediate products as well as the binary components are very reactive agents during the various stages of production. It would accordingly be constructive to seek to harness their reactivity for a more worthwhile endeavour—namely, for the manufacture of civilian products.

It may be argued that there is no need for exploring conversion possibilities since the civilian products can be manufactured more cheaply by the use of currently available techniques; furthermore, there is no economically justifiable application for the final and intermediate CW agents in the chemical industry. However, such assessments have been made in the absence of intensive studies and of basic and applied research in this field. Chemists and technical specialists employed in development centres for CW agents have not been encouraged to participate in investigations relating to the problem of conversion. Chemists in the civilian sector of industry have not been given adequate information on CW agents, and like their military counterparts, they have not been asked to deal with these questions. It should not be very difficult for competent representatives of military laboratories to meet with civilian chemists specializing in, say, pesticides or pharmaceuticals, to discuss conversion possibilities.

Although the conversion of CW agents cannot be considered the ultimate alternative to their destruction, it remains a worthwhile endeavour. It may lead to reconsideration of continuing the arms race in the field of chemical munitions and may arouse more concern for disarmament.

IV. Selected examples of the use of conversion products

Examples of substances related to technically significant CW base materials and intermediate products are included in the following list:

(*a*) *From the production of tear gases (CS, CN, CR)*
chlorobenzaldehyde
malononitrile
chloroacetyl chloride
diphenylamine
arsenic chloride
benzilic acid esters

(*b*) *From the production of vesicants of the mustard-gas type*
thiodiglycol
thionyl chloride
sulphuryl chloride
triethanolamine

(*c*) *From the production of nerve gases of the sarin/soman and the VX type*
phosphorus trichloride
phosphorus pentachloride
phosphorus sulphochloride
methylphosphonic chloride
methyl phosphates
alkyl- and dialkylamine compounds

Most of these compounds and most of the other intermediates of CW agents can find application in the civilian chemical industry for the manufacture of pesticides, pharmaceuticals, dyestuffs or plastics. These substances are valuable as base products and also in terms of the essential raw materials from which they are synthesized (Senning, 1971–72; CIBA, 1972; Fest and Schmidt, 1973; Oae, 1977).

Figure 1. New intermediates (based on mustard gas)

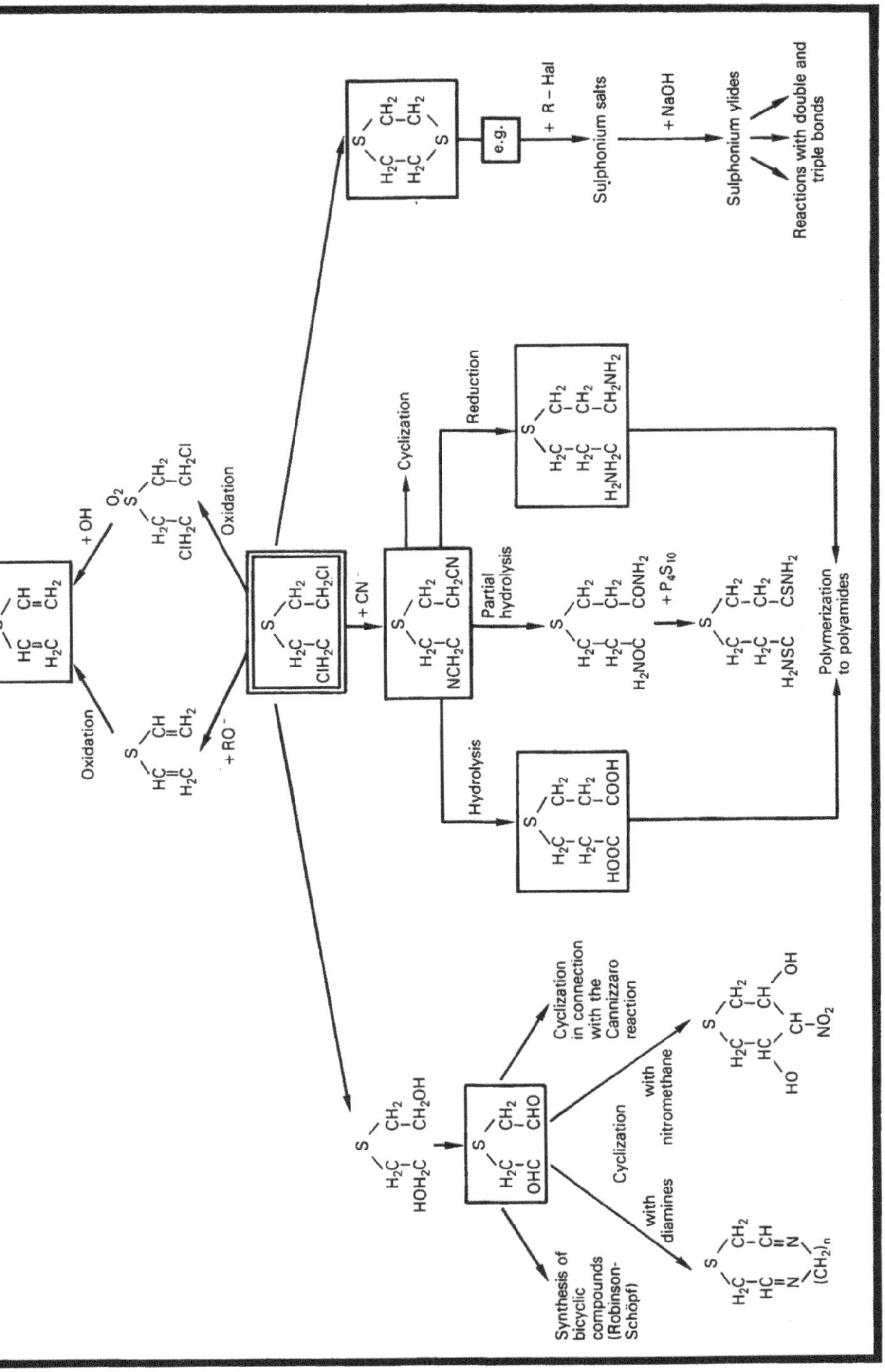

The conversion of CW agents or of their intermediate products is bound to lead to completely new synthetic components, which so far have been only of academic interest or have not at all been described in the literature. Chemists should look upon this work as a challenge, urging them on to perform syntheses with the help of these compounds and possibly leading them to hitherto unknown classes of substances.

Methods for obtaining new compounds from vesicants of the mustard-gas type, from several organophosphorus CW agents of the sarin/soman group and from the VX materials are shown in figures 1–5. Further examples and details regarding the reactivity of intermediate products of CW agents, especially organophosphorus compounds, are given in the literature (e.g., Lohs, 1974).

Figure 2. Total hydrolysis of G-agents to methanephosphonic acid as precursor for potential pesticides

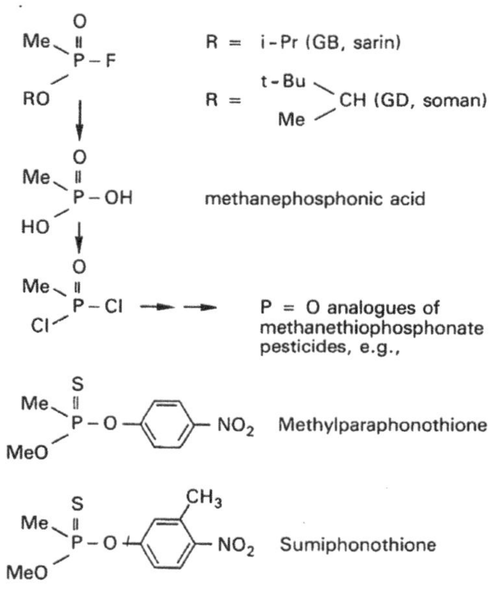

Figure 3. Chemical transformation of G-agents by nucleophilic replacement of the fluorine atom followed by substitution of the alkoxy group

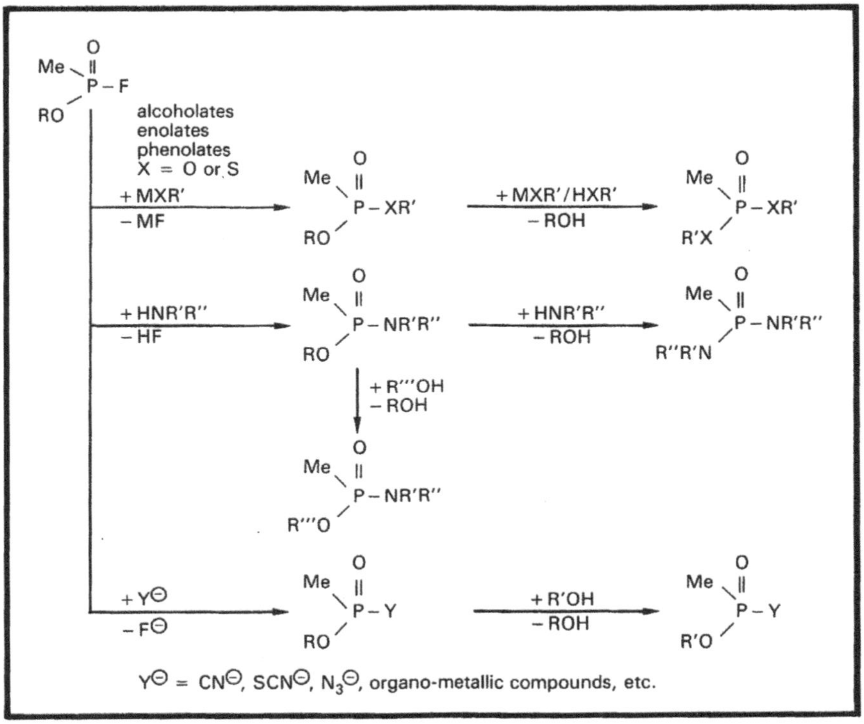

Figure 4. Chemical transformation of G-agents involving addition reactions

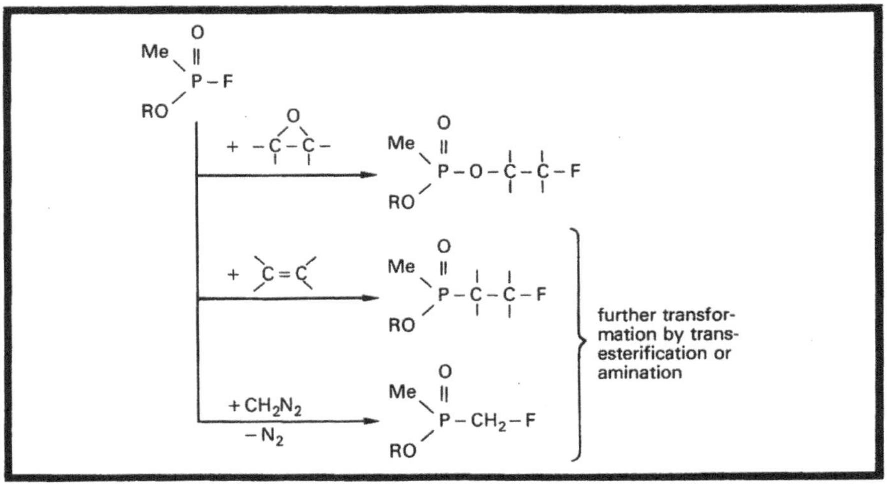

Figure 5. Chemical transformation of V-agents

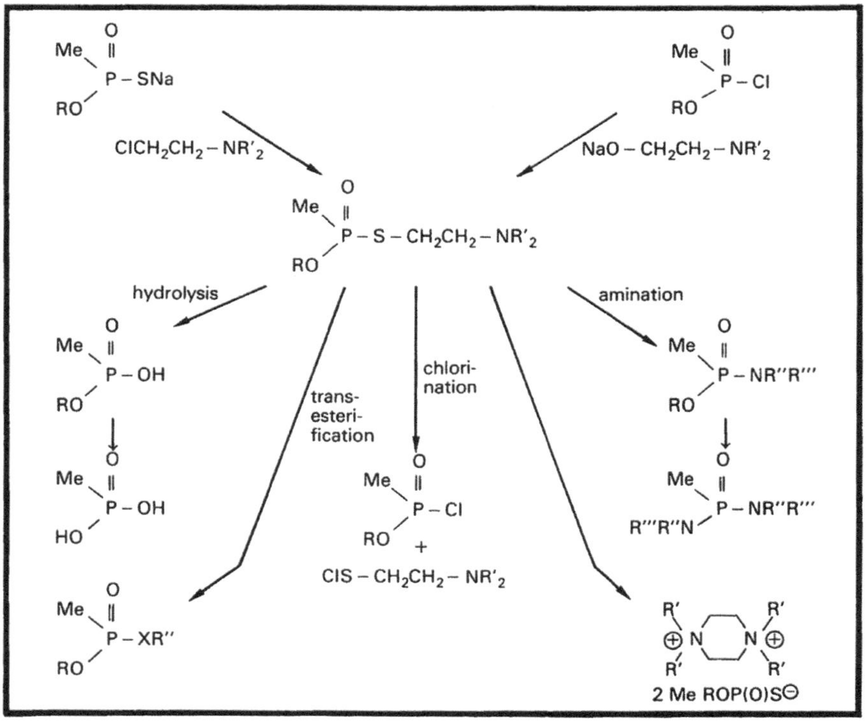

V. Conclusions

In view of the problems of conversion of CW agents and in view of their value as base materials and intermediate products, the following conclusions may be drawn from the disarmament standpoint:

1. Chemical warfare agents should not, as highly reactive compounds, be considered exclusively as final products. They can also be regarded as base materials and intermediate products for new chemical syntheses. Additional research and development is, however, necessary.

2. The base materials and intermediate products for the manufacture of CW agents are convertible into secondary products usable for civilian purposes in technological and engineering facilities in the chemical industry.

3. Special knowledge is required for handling CW agents. This knowledge is necessary for the production and further processing of these compounds, as well as for their detoxification or destruction. Specially skilled scientific and technical personnel can thus be kept fully engaged in work on destruction or conversion, and fears of redundancy are groundless.

4. The conversion of CW agents and of their base materials and intermediate products is, by and large, a technically feasible alternative to the unconditional destruction of these compounds.

5. Research in the field of conversion must be performed openly so as to provide a basis for confidence between states.

References

CIBA, 1972. *Carbon–Fluorine Compounds, Chemistry, Biochemistry and Biological Activities*, A CIBA Foundation Symposium. Amsterdam, London, New York: Elsevier/Excerpta Medica/North-Holland.

Fest, C. and Schmidt, K.-J., 1973. *The Chemistry of Organophosphorus Pesticides*. Berlin, Heidelberg, New York: Springer-Verlag.

GDR, 1976. Disarmament Conference document CCD/506, 6 July.

Lohs, Kh., 1974. *Synthetische Gifte*, 4th edition, pp. 200–24. Berlin: Militärverlag der Deutschen Demokratischen Republik.

Lohs, Kh., 1978. Verbot chemischer Kampfstoffe dringend erforderlich. *Deutsche Aussenpolitik*, **23**: 36–48.

Meissner, H. and Lohs, Kh. (eds), 1978. *Abrüstung, Wissenschaft, Verantwortung*. Berlin: Akademie-Verlag.

Meselson, M. (ed.), 1978. *Chemical Weapons and Chemical Arms Control*. New York and Washington, D.C.: Carnegie Endowment for International Peace.

Nerlich, U., 1977. Die Bedeutung chemischer Kampfmittel für die Verteidigungskonzeption 1980–90 aus der Sicht der Bundesrepublik Deutschland. *Europäische Wehrkunde*, **7**: 337–43.

Oae, S., 1977. *Organic Chemistry of Sulfur*. New York and London: Plenum Press.

Senning, A. (ed.), 1971–72. *Sulfur in Organic and Inorganic Chemistry*, vols 1–3. New York: Marcel Dekker.

SIPRI, 1975. *Chemical Disarmament: New Weapons for Old*. Stockholm: Almqvist & Wiksell.

SIPRI, 1978. The destruction of chemical warfare agents. In *World Armaments and Disarmament: SIPRI Yearbook 1978*, pp. 360–76. London: Taylor & Francis.

SIPRI, 1979. Stockpiles of chemical weapons and their destruction. In *World Armaments and Disarmament: SIPRI Yearbook 1979*, pp. 470–89. London: Taylor & Francis.

World Federation of Scientific Workers, 1977. *Ending the Arms Race—The Role of the Scientist*. London.

Lessons learned from the destruction of the chemical weapons of the Japanese Imperial Forces

H. KURATA

The Joint Staff College of Defense Agency, Tokyo, Japan

Abstract. Chemical weapons are inhumane weapons of mass destruction. In addition, they could be used by terrorists as weapons of coercion. There is, accordingly, an urgent need for concluding a treaty prohibiting these weapons. The treaty should prescribe destruction methods that are thorough, safe and secure. There is a need for establishing an optimal elimination system employing operational methods that would serve to prevent disastrous accidents, such as those experienced by Japan after World War II. This should be accomplished via scientific study by an international group of experts possessing a wide and integrated perspective.

This paper first discusses the chemical warfare potential of the Japanese Imperial Forces at the end of World War II, the operations for the elimination of chemical weapons performed after the war, and the accidents that occurred during the post-war period as a result of the defective nature of these elimination operations. The causes of the accidents are examined, and a number of reflections on the Japanese experience are presented. These considerations are then incorporated in recommendations for future chemical weapons destruction systems under a treaty prohibiting chemical weapons.

I. Introduction

The Japanese Imperial Forces existed from the year 1872 up to August 1945. The development of the armed forces in Japan during this period was guided by the slogan "Catch up with and outstrip the developed nations of Western Europe". After World War I, Japan ceased copying European weapons and began designing, developing and testing its own weapon systems, particularly in the field of chemical weapons.

The dispute with the Soviet government in Siberia in 1919 gave Japan its first opportunity for developing a chemical warfare (CW) potential. This CW capability reached its peak around 1940; it was not, however, used at all in World War II.

At the time of Japanese disarmament, most of the chemical weapons were handed over to the US occupation force and were eliminated under its supervision. The elimination operations were defective, however, which later led to numerous accidents: a total of 102 accidents, resulting in 127 casualties and 4 deaths, have occurred since World War II in connection with the chemical weapons elimination efforts in Japan.

It was found that the accidents were due mainly to a combination of two factors: (*a*) the inadequate disposal operations previously carried out by the Japanese Imperial Forces, and (*b*) a lack of information about this work of destruction—including the whereabouts of the weapons—since the pertinent documents had been burnt or lost. The destruction operations performed under US military supervision were also defective.

In order to investigate these accidents and to focus concern on locating the residual inadequately destroyed chemical weapons, the Prime Minister of Japan in 1972 ordered an overall national inspection of chemical weapons of the former Japanese Imperial Forces.

The aim of this paper is, first, to describe the Japanese experience of chemical weapons destruction in the hope that similar problems and complications can be avoided in the future. The paper traces the development of the Japanese Imperial Forces, and describes the chemical weapons elimination operations at the end of World War II and also the accidents and re-elimination operations entailed during the post-war period. Second, on the basis of the lessons learned from the Japanese experience, the features necessary for devising an optimal chemical weapons elimination system are presented. These features could be applied in a future chemical disarmament treaty.

II. Chemical warfare potential of the Japanese Imperial Forces at the end of World War II

Development of chemical warfare capabilities

In his book *Chemical Warfare*, the German author Hanzurian wrote:

> ...the Japanese Forces are sure to have developed the same chemical warfare capabilities as the great powers.... No information is available on Japan's military chemical technology. The country has no specialized chemical units, and it is accordingly difficult to obtain any information on the size of its chemical warfare efforts.
>
> There is a modern research institute in Tokyo that is studying problems of military chemistry, chemical warfare agents and chemical weapons. Two factories produce the chemicals necessary for peace-time use by units. There are four large government-owned factories located in Tokyo, Osaka, Nagoya and Kokura, which we can assume are capable of producing chemical shells for wartime use. There are also four naval arsenals located at Yokosuka, Kure, Hiroshima and Sasebo. (Hanzurian, 1943)

The above account covered almost all the CW research, production and storage facilities of the Japanese Imperial Forces towards the end of World War II. When the chemical instruction facility, the Narashino School, is added, the picture is complete. The research institute cited by Hanzurian was the No. 6 Army Research Institute in Tokyo, and the factories were the Chukai

Figure 1. Location of chemical weapons facilities of the Japanese Imperial Forces at the end of World War II

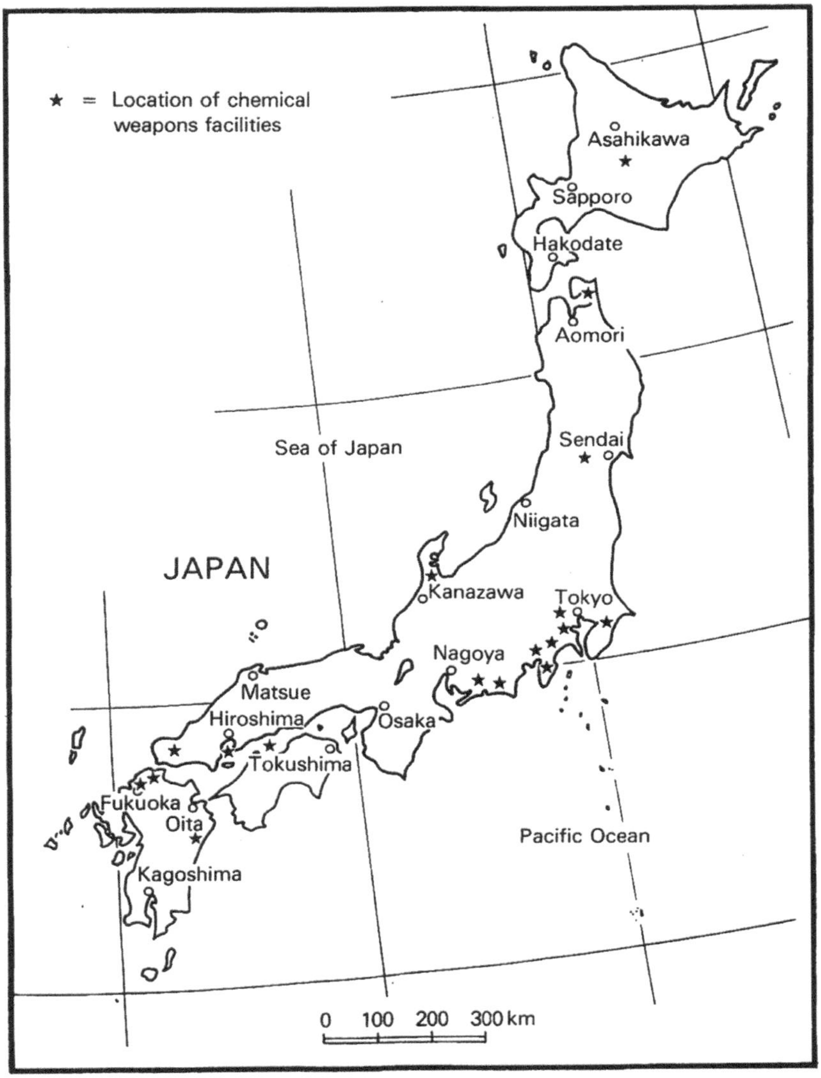

Factory on Okino-shima Island in the Seto Inland Sea and the Sone Factory in the northern part of Kyushu (Tani, 1955). Figure 1 shows the locations of the chemical weapons facilities of the Japanese Imperial Forces at the end of World War II. The Narashino School was later converted to the Narashino Camp of the Japan Ground Self-Defense Force (JGSDF), and the No. 6 Army Research Institute to a collective housing area. The whole of Okino-shima Island was converted to a recreation centre, and the Sone Factory rebuilt as a JGSDF camp.

Chemical weapons status at the end of the Pacific War (1942–45)

Precise information on the chemical weapons status of the Japanese Imperial Forces is unattainable since virtually all the pertinent documents had been burnt or were lost. From the many post-war accidents that occurred, it may be inferred, however, that many of the chemical weapons elimination operations were defective. The accidents pointed to the need for a thorough inquiry into chemical weapons in Japan.

On 24 May 1972, during a session of the Diet, the Opposition requested the Prime Minister to conduct an inquiry into the status of chemical weapons. The Prime Minister acceded to the request and appointed the Environmental Agency in charge of an investigation into the production and disposal of poison-gas munitions (Japanese Diet, 1972). The inquiry report was completed in July of the same year (Environmental Agency, 1972).

Table 1. Standard chemical warfare agents and chemical munitions of the Japanese Imperial Forces

Chemical warfare agents

Physiological classification	Standard name	Year of standardization	Common/chemical name
Lung irritant	Blue No. 1	1931	Phosgene
Lachrymator	Green No. 1	1931	Bromobenzyl chloroacetophenone
	Green No. 2	1931	
Vesicant	Yellow No. 1A	1931	Mustard A[a]
	Yellow No. 1B	1931	Mustard B[b]
	Yellow No. 1C	1936	Mustard + non-freezer
	Yellow No. 2	1931	Lewisite
Sternutator	Red No. 1	1931	Diphenyl cyanarsine
Blood gas	Brown No. 1	1937	Hydrogen cyanide
Smoke	White No. 1	1931	Trichloroarsine

Chemical munitions

Type of gun	Shell designation	Year of standardization	Filling agent(s)
7-mm how	Blue + White	1932	Blue No. 1 + White No. 1
	Yellow	1932	Yellow No. 1
10-mm how & can	Red	1933	Red No. 1
15-mm how & can	Brown	1937	Brown No. 1
Light mortar	Yellow	1935	Yellow No. 1
	Red	1935	Red No. 1
	Heavy Blue	(1939)	Blue No. 1
	Heavy Brown	(1939)	Brown No. 1
Bomb	15-kg Red	Not known	Red No. 1
	50-kg Incendiary	1939	Incendiary agent

Abbreviations and conventions: "how" stands for "howitzer"; "can" stands for "cannon"; and "( )" denotes "test stage and not standardized".

Notes:

[a] Ethyl chlorohydrin + sodium sulphide + hydrogen chloride.

[b] Ethylene + sulphur chloride.

Table 1 shows the summarized results of the national inquiry as well as other data on CW agents and chemical munitions of the Japanese Imperial Forces (Japanese Imperial Chemical Society, 1938; Hisamura *et al.*, 1956).

The manufacture of chemical munitions involved CW-agent production at the Chukai Factory, followed by transport and loading into shells at the Sone Factory. The yearly production rates reached and maintained a peak from the time of the Chinese Incident (1937) to the middle of the Pacific War (1943). The production capacity during this period appears to have been about 6 400 tons per year. A breakdown by agent is shown in table 2 (Kawakami, 1965).

Table 2. Annual CW-agent production in Japan during World War II

Name of agent	Production in 1943 (tons)
Red	950
Yellow	3 000
Brown	2 100
Blue	320
Total	**6 370**

Table 3. Amounts of CW agents and chemical munitions in the possession of the Japanese Imperial Forces at the end of World War II[a]

Service	Agents (tons)				Munitions[b]	
	Blister[c]	Blood	Choking	Tear	Shells	Canisters
Army	2 278.6	13.2	958.1	7.0	116 000	573 642
Navy	446.7		100.9	35.5	2 000	
Not known[d]		976.8				
Total	**2 795.3**	**990.0**	**1 059.0**	**42.5**	**118 000**	**573 642**[e]
Agent total		**4 886.8**				

There was evidence of other chemical weapons, but types and amounts are not known.

Notes:

[a] This is a summary of the amounts from the reports submitted to the US force and the record of elimination.

[b] It is difficult to convert from shell content to amount of agent.

[c] This is a mixed sum of (mostly) mustard and (some) lewisite.

[d] This is the difference between the amount destroyed and that possessed at the end of World War II.

[e] The amounts destroyed were not recorded.

Around 1944 the plight of Japan in World War II became increasingly difficult, however, and the production of chemical weapons was stopped. The factories manufacturing chemical weapons gradually changed over to the production of explosive compounds and incendiary shells. This fact is borne out by a policy statement issued by the Japanese government, notifying the Allied powers through Switzerland that "Japan shall not use gas weapons in this war" (Ministry of Foreign Affairs, 1944).

There were two reasons for ceasing production of chemical weapons (Akiyama, 1955). First, it was believed that the Allied powers would not use chemical weapons during the war. Nor did Japan intend using them. If, however, it later turned out that chemical weapons were used, Japan had sufficiently large stockpiles for retaliatory use. Second, it seemed more realistic to maximize the production of explosive compounds and incendiary shells than to produce chemical rounds which were likely to remain unused.

On the basis of the facts given above, and of the national inquiry report, it is possible to estimate the total quantities of CW agents and chemical munitions which the Japanese Imperial Forces possessed at the end of World War II. These amounted to approximately 4 900 tons of CW agents, 118 000 shells and 574 000 canisters (table 3).

III. Chemical weapons destruction operations at the end of World War II

Owing to the lack of official records, details of the operations for the elimination of chemical weapons at the end of World War II were not known. Nor was the status of chemical weapons known up to the time of the above-mentioned national inquiry of 1972. However, piecing together information from the inquiry report, from some undestroyed documents and from the testimony of various people involved, it is possible to state that the elimination operations were conducted in the following way (Environmental Agency, 1972):

1. Most of the operations for the elimination of chemical weapons and CW agents were carried out under orders from the US occupation force in Japan and under its supervision. US forces supervised the operations from ships or from aircraft (such as the OY–1 liaison plane). The main disposal areas assigned by the US forces are shown in figure 2.
2. The methods deemed the safest and the most reliable were destruction by burning and disposal by dumping in the sea. The latter method was adopted.
3. The standard laid down for the sea-dumping areas was that they should be more than 10 nautical miles (*c*. 18.5 km) distant from the shore and more than 3 000 feet (*c*. 1 000 m) deep.
4. The dumping operations were carried out by Japanese workers using chartered disposal ships.

Figure 2. Sea areas assigned for dumping chemical weapons at the end of World War II

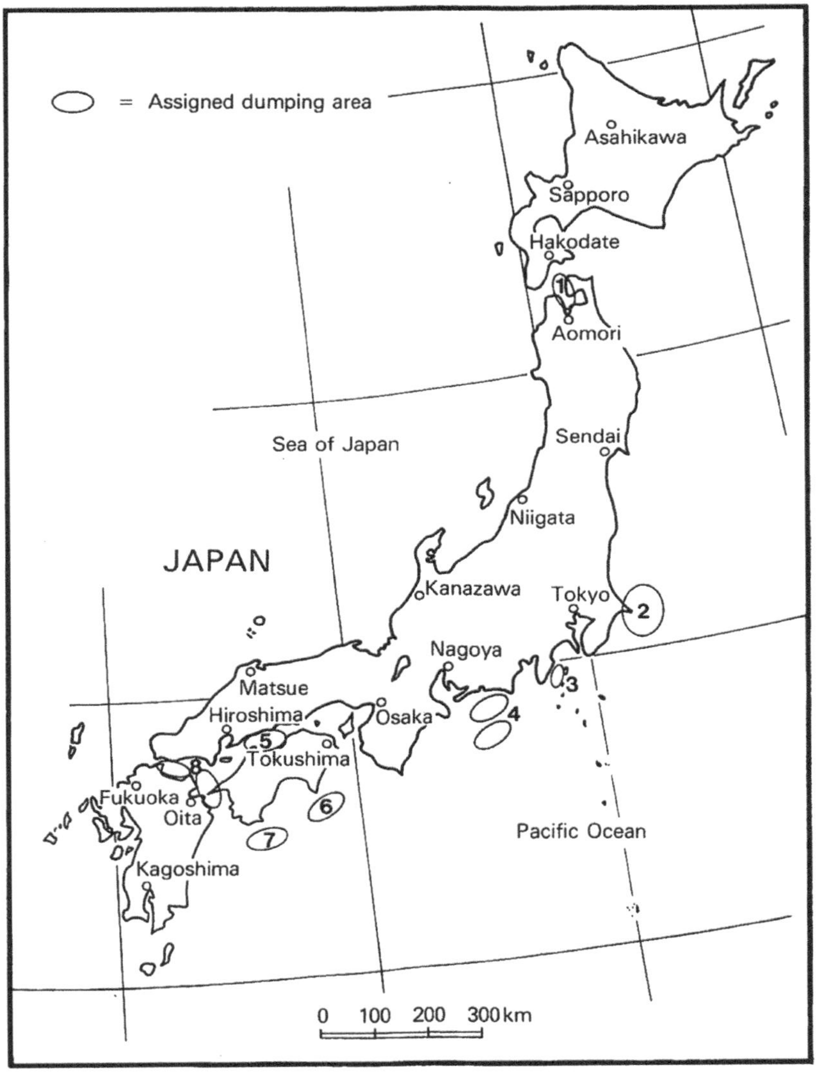

IV. Recurring problems of defectively destroyed chemical weapons

Accidents and their causes

Most of the accidents occurred at the end of World War II in the vicinity of the main dumping areas and were due to incomplete destruction. However, many accidents also occurred at unexpected places away from the main dumping areas, indicating that, unknown to the US forces, other incomplete elimination operations had been carried out privately by the Japanese Imperial Forces at the end of World War II. Such accidents occurred throughout Japan—there were 102 cases, 20 of which resulted in casualties, including 4 deaths and 123 wounded. The chronology of accidents given in figure 3 shows that accidents

Figure 3. Chronology of accidents and casualties caused by defectively destroyed chemical weapons

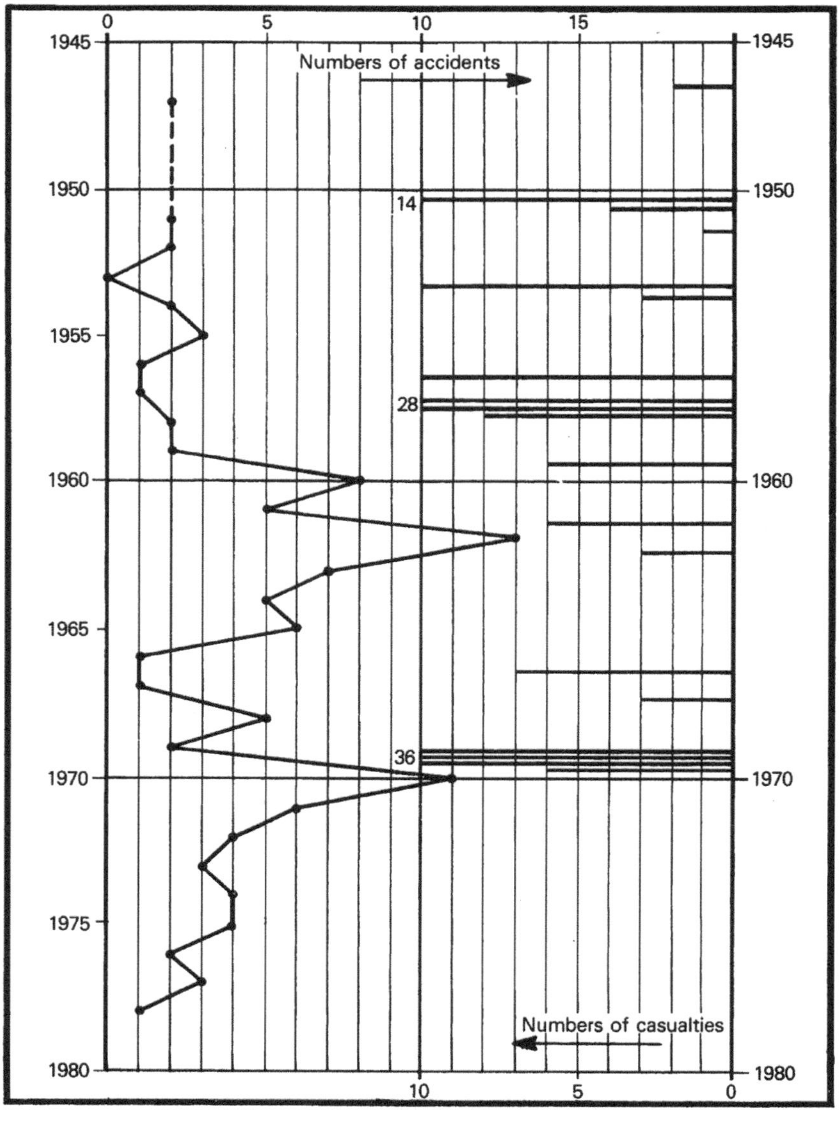

Figure 4. Areal distribution of accidents/casualities caused by defectively destroyed chemical weapons

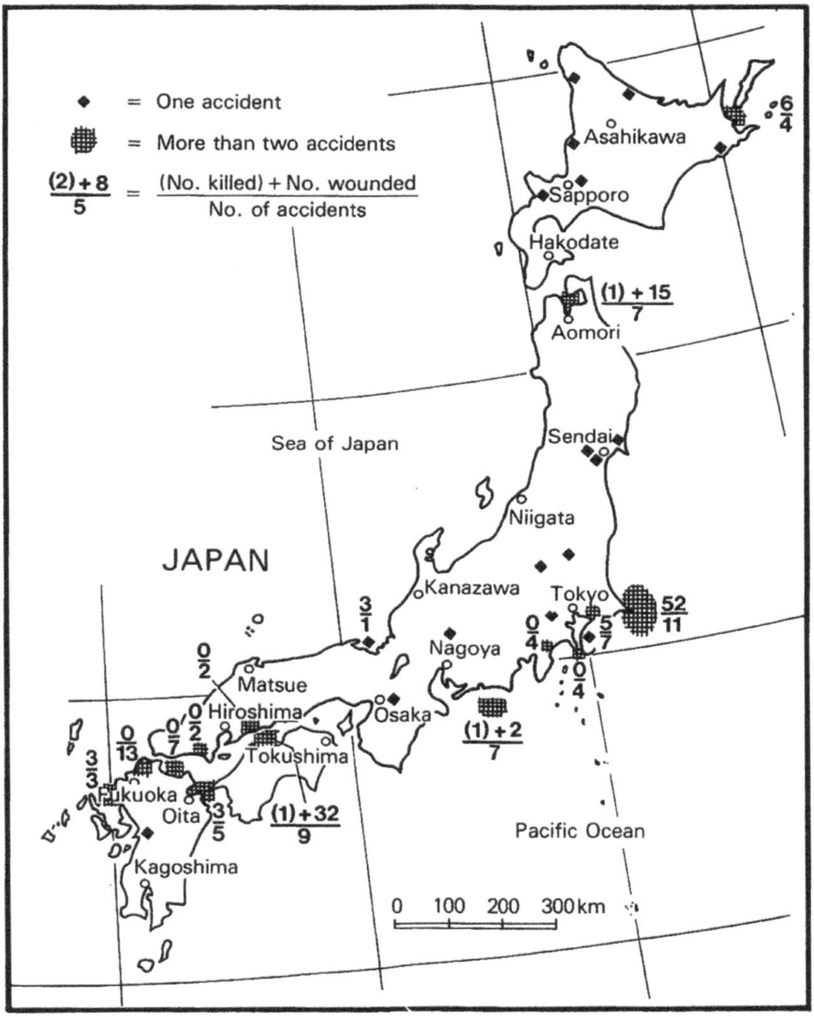

have occurred since 1971; fortunately, these accidents did not lead to any casualties. Figure 4 shows the areal distribution of accidents and casualties caused by defectively destroyed chemical weapons.

In his investigations of the causes of accidents that had led to casualties, the author of this paper found that the victims had often approached the incompletely destroyed chemical weapons unwittingly—for example, fishermen finding such weapons caught in their nets, or workers digging them up at construction sites for buildings and roads. Unfortunately, accidents also occurred during the redisposal operations, when several members of the Japan Self-Defense Force (JSDF) were injured by explosions of previously defectively destroyed chemical weapons.

Reflections on the elimination operations

The main problems in the chemical weapons elimination operations at the end of World War II were as follows:

1. Certain quantities of chemical weapons were not handed over to the US forces. They were destroyed privately and, as it turned out, defectively. This becomes evident by overlaying figure 2 (disposal areas) on figure 4 (distribution of accidents). The composite is shown in figure 5.

Figure 5. Composite of figure 2 (dumping areas) and figure 4 (distribution of accidents)

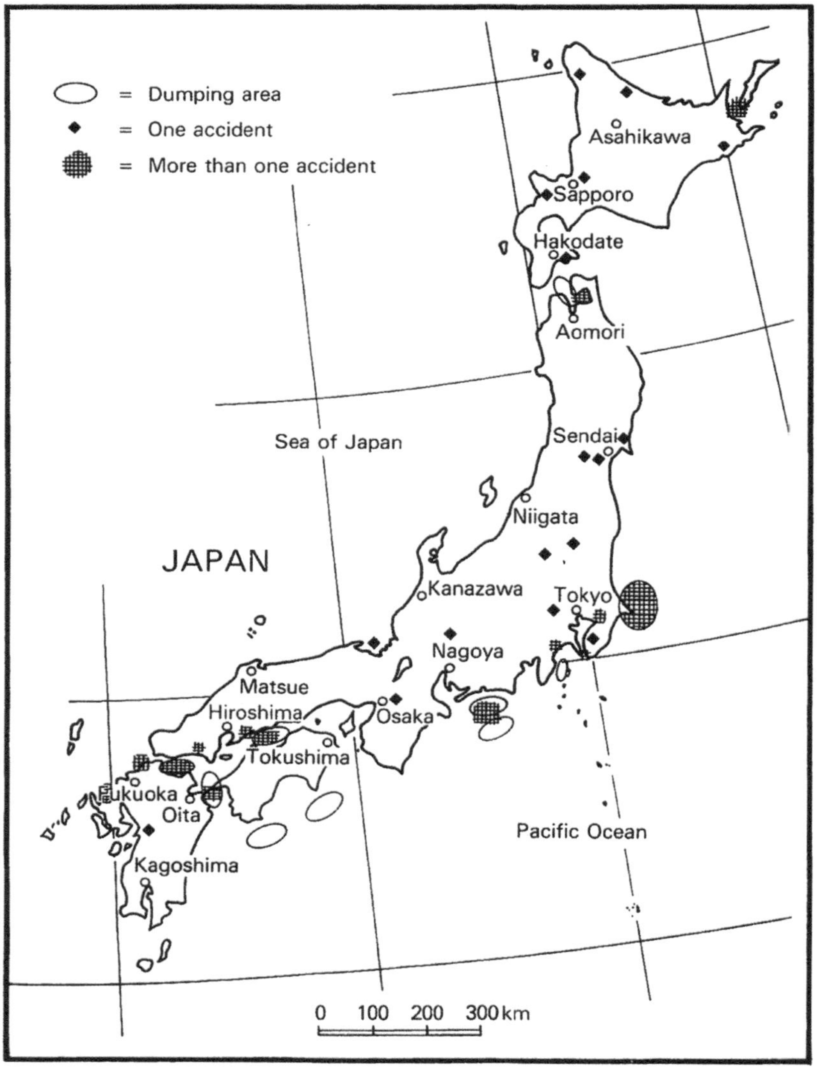

2. The sea-dumping areas were designated without a prior study of such factors as coastal fishing and environmental pollution. Moreover, the dumping operations grossly violated the specifications set for the assigned areas. As shown in figure 6, the standard sea-dumping areas were to be more than 10 nautical miles (*c*. 18.5 km) distant from the shore and more than 3 000 feet (*c*. 1 000 m) deep. However, by examining the locations of the main dumping areas relative to these criteria, it becomes clear that the prescribed standard was often violated (figure 7).

Figure 6. Contour line at 1 000 metres sea-depth around Japan

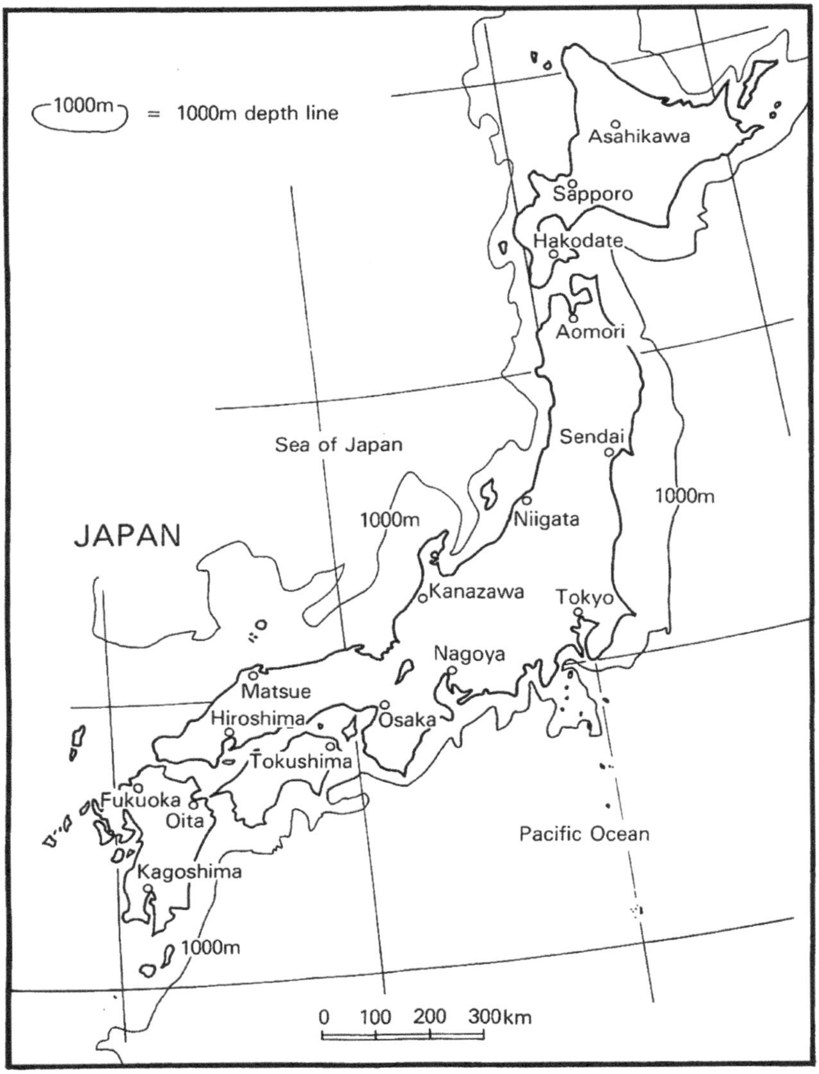

Figure 7. Composite of figure 2 (assigned dumping areas) and figure 6 (1 000 m depth line)

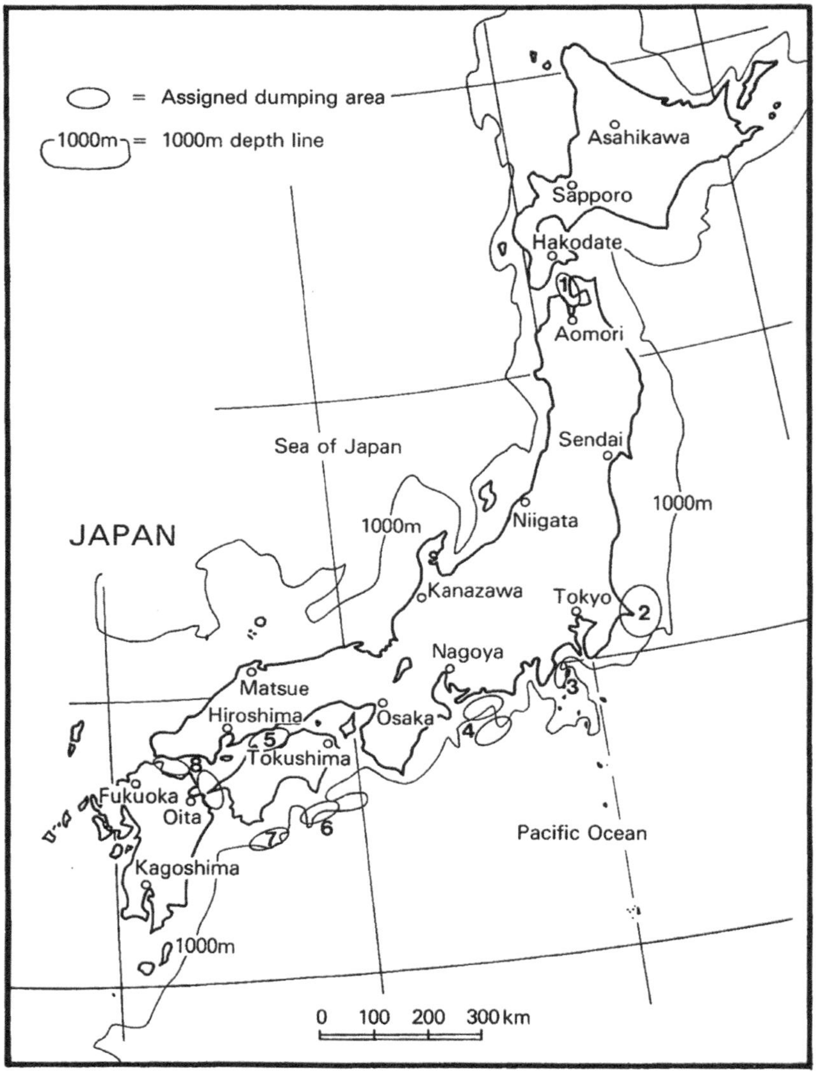

3. The sea dumping areas were too near the shore. If those who were responsible for planning and implementing the dumping operations had borne in mind the fact that coastal fishing supplied the primary food resources of the Japanese people, the dumping areas might have been shifted beyond the continental shelf. However, it may have been that, since the operation of removal of sea-mines laid by the Allied powers was still not complete, dumping at the assigned locations (figure 2) was judged hazardous.

4. The dumping operation itself was faulty. CW agents in drums and munitions without fuses were transported to the sea areas in question and dumped. The dumping areas were shallow enough for the oxygen content of the water to erode the metal drums. The possibility of the release of chemical agents caused by metallic erosion and rough handling of containers and shells, with its attendant dangers, should have been taken into account. Most of the accidents caused by catching CW materials in fishing nets could have been avoided if the standard laid down by the US force had been observed, namely, that sea-dumping was to be performed beyond the continental shelf.

5. As figure 3 shows, there were two peaks of accidents during 1950–70, the post-war period before operations for re-eliminating chemical weapons were begun. If the national inquiry into chemical weapons had been conducted in the 1950s rather than in 1972 (i.e., 27 years after the end of World War II) and if the facts concerning the incomplete disposal of CW agents had been well known, the number of accidents that caused injuries could have been considerably reduced. In addition, an earlier inquiry would have yielded more substantial results, since a greater number of those who had been responsible for chemical weapons in the Japanese Imperial Forces would have been available to testify.

6. Finally, the most serious problem was the incompleteness of pertinent documents. Greater attention should have been paid to monitoring and documenting the status of chemical weapons at the end of the war since they were classified weapons, possessing a great potential for harm if not thoroughly investigated and carefully destroyed.

V. Lessons to be learned

The following lessons are to be learned from the disposal of CW agents and chemical munitions by the Japanese Imperial Forces at the end of the Pacific War and from the disastrous post-war accidents caused by defective disposal operations. These lessons should be useful for the future.

1. *Reliable elimination/re-elimination operations.* Reliable elimination and re-elimination methods for the various kinds of chemical weapon should be developed so as to ensure that the operations are performed completely, safely and securely. There should be proper supervision of the operations, and a system for recording and retaining all documents should be devised.

2. *Anti-accident measures.* If chemical weapons were disposed of correctly, accidents would be less likely to occur. However, in addition to human carelessness, natural calamities—such as earthquakes or big fires—can also lead to accidents. Thus, measures for dealing with unforeseen accidents should also be established.

3. *Inspection of the disposal process.* Defective inspection at every stage of chemical weapons disposal was a causative factor of the accidents in Japan.

The various stages of a disposal system, such as reception of a consignment of weapons from the possessor country, transportation, storage, actual disposal operations, recording and notification, should be clearly set out in a future treaty prohibiting chemical weapons. There should be proper inspection at every stage.

4. *Retention of records of chemical weapons status*. Information on chemical weapons is usually classified. In view of the nature of these weapons, all detailed documents specifying the types of weapon, method of manufacture, toxicity, place of storage, and so on, should be submitted to an international institute. Subsystem files of documents on each step of the disposal system should also be developed and retained.

In view of the present international situation, the following factors should also be taken into account:

(*a*) *Safety and security measures*. The present international situation makes it necessary to take stringent safety and security measures at each step of the destruction process for chemical weapons. It is particularly important to guard against theft or capture of the weapons by terrorists during the transportation and storage phases.

(*b*) *Dissemination of data*. Data on the types of chemical weapon, quantities, transportation and storage status, and current status of disposal operations by the disposal agency should be made public to all countries. This may help to engender confidence between countries and thus contribute to world peace. It would also contribute to safety and security during transportation and storage.

The above points show that it is necessary to make a thorough study now of the methods and procedures for the disposal of chemical weapons so as to attain an optimal disposal system.

VI. Recommendations for future chemical weapons destruction systems

Elimination of incompletely destroyed chemical weapons

Prompt and reliable re-elimination of incompletely destroyed chemical weapons should be carried out so as to minimize the likelihood of occurrence of (further) accidents. In the post-war period Japan has been responsible for carrying out the destruction operations in all the 102 cases of accidents. A system and a methodology for chemical weapons elimination have been established in Japan on the basis of the experience gained. The methods in use today consist in neutralization and incineration or neutralization and sea-dumping. The actual method used depends on the kinds and amounts of chemical weapons to be eliminated and on the extent of erosion/leakage of the containers and shells.

Figure 8. Flow diagram showing Japan's present system for re-eliminating chemical weapons

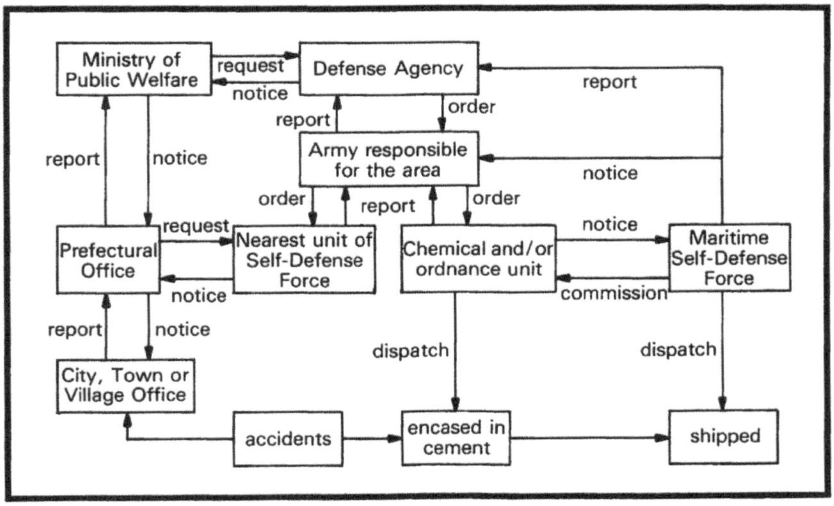

Japan's present system for chemical weapons re-elimination is shown diagrammatically in figure 8. When any chemical weapons are discovered, the Prefectural Office is notified via the City, Town or Village Office. The Governor then requests the dispatch of GSDF demolition experts through the nearest JSDF camp. The Army dispatches a crew of the chemical corps[1] in the case of CW agents only, and an officer of the chemical corps and a crew of the ordnance corps in the case of chemical munitions. Elimination operations are carried out by the crew concerned and by civilians under the supervision of a chemical officer. The items are decontaminated (neutralized) and placed in 50-gallon (*c.* 190-litre) drums, which are then encased in cement. The encased drums are then entrusted to Maritime Self-Defense Force (MSDF) units, who dump them in the sea beyond the continental shelf at a depth of over 1 000 metres so as to ensure effective disposal.

Criteria of an effective destruction system for chemical weapons

Among the criteria which must be considered in designing an effective destruction system for chemical weapons are: ownership of the system, scope of the system, safekeeping of documents, safety control, and security of the system. The following alternative destruction systems are given by way of

[1] The Ground Self-Defense Force has a chemical corps responsible for NBC (nuclear, biological and chemical) defence. The corps is very small, having only about 200 officers.

illustration with respect to the first criterion listed above—namely, ownership of the system:

1. *Destruction of chemical weapons in the country where they are stationed.* The weapons are destroyed (*a*) by the host country, the results being notified to an international institute, or (*b*) under the surveillance of supervisors from an international institute.

2. *Destruction of chemical weapons by an international institute.* The weapons are destroyed (*c*) at an elimination site constructed by the international institute, or (*d*) using the elimination sites of other countries.

3. *Destruction of chemical weapons by some other country in co-operation with an international institute.* (*e*) The destruction is supervised by the international institute, or (*f*) the destruction records are sent to the international institute.

The above six alternatives (*a*)–(*f*) are the main candidate systems and each will have several subsystems. As illustrated by the example above, several systems should be considered for each criterion of the destruction system. These matters should be taken up for systematic discussion by research committees or groups appointed by the UN General Assembly or the Committee on Disarmament. It is to be hoped that the lessons learned from Japan's experience will then be given close attention.

VII. Conclusions

Chemical weapons, like nuclear weapons, are weapons of mass destruction. They also lend themselves to capture and use by terrorists as weapons of coercion.

The prohibition of the use of chemical weapons was agreed to in the Geneva Protocol of 1925 (concluded as a result of the experiences of World War I). Since that time, chemical weapons have been used only on a few occasions in wars and conflicts. However, the Protocol does not prohibit research and development or production and stockpiling in the case of chemical weapons. Hence the need for concluding a chemical disarmament treaty is urgent.

Based on a thorough scientific study by experts having a broad perspective, the completeness, safety and security of elimination methods for stockpiled chemical weapons should be examined, and an optimal system established.

The Japanese Imperial Forces were equipped only with the types of chemical weapon used in World War I, but today certain CW-capable countries have in their arsenals much more modern chemical agents, such as the nerve gases, which are several times more toxic than the agents dating back to the time of World War II. The results of accidents caused by the incomplete disposal of modern chemical weapons defy the imagination. Thus, the importance of the certain, safe and secure disposal of CW agents and chemical munitions cannot be overemphasized.

References

Akiyama, T., 1955. *Research History of Chemical Weapons in No. 6 Imperial Army Research Institute.* Tokyo: Ministry of Public Welfare (in Japanese).

Environmental Agency, 1972. National investigation on poison-gas munitions etc. of the Japanese Imperial Forces. Environmental Agency Report (in Japanese).

Hanzurian, R., 1943. No. 6 Imperial Army Research Institute. In: [Chemical warfare], p. 261. Tokyo (in Japanese; translated from the German).

Hisamura, T. *et al.*, 1956. *History of Chemical Weapons of Imperial Japan.* Tokyo: Ministry of Public Welfare (in Japanese).

Japanese Diet, 1972. Proceedings of No. 68 Diet, the Settlement of Accounts Committee of the House of Councillors. *The Official Gazette*, 24 May 1972. Tokyo (in Japanese).

Japanese Imperial Chemical Society, 1938. *Handout of a Course of Gas Weapons Defence.* Tokyo: Japan Pharmacist Society Press (in Japanese).

Kawakami, M., 1965. Outline of the Japanese Imperial Army Sone Arsenal. *Journal of Chemical School of JGSDF*, **10**: 36–43 (in Japanese).

Ministry of Foreign Affairs, 1944. The question of the prohibition of the use of war gas. In *Precedents of Signature and Ratification of Treaties*, vol. 3, August. Tokyo (in Japanese).

Tani, I., 1955. *Outline of the Chemical Warfare System of the Japanese Imperial Army.* Tokyo: Ministry of Public Welfare (in Japanese).

Some toxicological problems in the destruction of chemical warfare agents

V. VOJVODIĆ
Military Medical Academy, Belgrade, Yugoslavia

and Z. BINENFELD
University of Zagreb, Faculty of Natural Sciences & Mathematics, Zagreb, Yugoslavia

Abstract. Chemical warfare agents (including binary compounds) are classified in different categories on the basis of various criteria. Some methods for their destruction and the possible attendant hazards are mentioned. The most important toxicity parameters and tolerance levels are discussed, and their significance for human safety in the working atmosphere is outlined, especially in connection with destruction processes for chemical warfare agents. The tasks of the medical services in obviating or reducing the risk of poisoning incurred by personnel and in providing first aid and treatment are indicated, and some recommendations made.

I. Introduction

Classification of chemical warfare agents

All chemical compounds, when viewed for their possible use as chemical warfare (CW) agents, may be divided into the following categories: (*a*) single-purpose CW agents, (*b*) dual-purpose CW agents, and (*c*) single-purpose non-CW agents.

Binary chemical munitions may contain (*a*) single-purpose binary components, (*b*) dual-purpose binary components, or (*c*) a combination of both (Yugoslavia, 1976).

All chemicals used as CW agents or needed for their production may be divided into the following categories:

(*a*) *Raw materials*. These are the starting substances for the synthesis of war gases—for example, yellow phosphorus, sulphur monochloride, ethane, hydrofluoric acid, and so on (SIPRI, 1971; Yugoslavia, 1972).

(*b*) *Intermediates*. This group includes all compounds which react with one or more compounds to produce a CW agent.

(*c*) *Final products*. These are classified CW agents.

CW agents can be stockpiled in bulk or filled in various types of munition, such as (*a*) containers in which the CW agents are relatively easily accessible; (*b*) munitions containing explosives (and perhaps propellants) that are not admixed with CW agent, and where access to the CW agent is accordingly also relatively easy; and (*c*) munitions in which the explosive component has to be defused before extraction of the CW agent. In the third case, the explosive component of the munitions poses special problems in the work of destruction.

Methods for the destruction of CW agents

An indispensable requirement of an international agreement on chemical disarmament is the destruction of stockpiles of CW agents and chemical munitions.

The literature contains relatively few papers on the destruction of CW agents. These papers comprise mostly CCD (Conference of the Committee on Disarmament) papers or US Army technical reports, dealing with the nerve agents, mustard gas and CS (USA, 1972; Canada, 1974; USA, 1974 *b*, 1974 *c*; SIPRI, 1975 *a*; Belcher, 1977; NTIS, 1977, 1978). In addition to prescribing the destruction of CW agents proper, a future chemical disarmament agreement should also prescribe the destruction of binary munitions, binary precursors and all single-purpose chemicals used for the production of CW agents. This requirement would, of course, add to the toxicological problems encountered during the destruction process.

Chemical destruction of CW agents generally yields non-toxic end-products. However, the nature of these products makes them hazardous for the environment if they are introduced directly into ground cavities, lakes or rivers. The disposal of large quantities of final products derived from the various chemical destruction processes requires special measures (SIPRI, 1975 *a*, 1978, 1979).

Pyrolysis of CW agents in autoclaves is technically feasible, but it may lead to complex by-products, about which little is known. Incineration of CW agents, with scrubbing of the exhaust gases for removal of environmental pollutants, appears to be a very promising method for the disposal of the mustards and nerve agents, both from the technical and environmental points of view.

The choice of method for the complete destruction and detoxification of CW agents should obviously take into account such factors as agent specificity, economic considerations, and personnel and environmental protection.

Hazards

The destruction of stockpiles of CW agents is a long-drawn-out process entailing environmental and other hazards. Rigorous procedures must be followed in destruction operations (USA, 1972; Canada, 1974; USA, 1974 *c*).

The process of destruction may be performed near the location of the stockpiles or at a distance from them. In the latter case, appropriate means of transportation for the bulk chemicals of munitions to be destroyed will be necessary.

Storage and destruction problems vary with the type of CW agent involved. Some of the CW agents used in World War I—such as chlorine, phosgene and hydrogen cyanide—require less stringent precautions than mustard gas or nerve gas. However, the basic precautions for health safety and against leakage are similar.

From the toxicological point of view, the danger in the course of destruction of CW agents arises from : (*a*) the substance(s) to be destroyed; (*b*) all the substances entering into the destruction process; (*c*) all the products formed during the destruction process; and (*d*) the final product(s) of the destruction process. The chemical substances entering into the destruction process, as well as those formed as intermediates or final products, may be gases, liquids or solids, depending on the type of process (incineration, chemical neutralization, catalytic process, and so on). Depending on the physical state of these substances, as well as on their physico-chemical properties, all kinds of toxication (acute, subacute, chronic or delayed) and all methods of penetration into the body (inhalation, percutaneous absorption or ingestion) are possible.

II. Toxicological parameters

Toxicity

Any assessment of the safety measures necessary to prevent contact with hazardous and toxic material invariably uses the term "toxicity" as the primary criterion, and the term "tolerance level" as the secondary criterion.

The *toxicity* of any material—natural or synthetic—is an expression of the quantity of that material which will kill or injure a living organism. It is important to appreciate that there is no sharp dividing line between lethal (killing) and non-lethal (injuring) doses; the probability of causing death gradually increases as the dose increases. In consequence, the sole information often published is that a particular dose may kill or incapacitate (injure). This dose may be described as the *lethal dose* or the *incapacitating dose*, but it is often impossible to determine with any precision what proportion among toxicated people would be killed or incapacitated. This is, of course, by no means specific to CW agents. Toxicity can be measured in a number of ways, depending on the route by which the material enters the body: through the nose (inhalation), through the mouth (ingestion) or through the skin (percutaneous absorption). In view of the different physico-chemical and physiological properties of the various CW agents, the following categories are suggested for use (USA, 1974*a*; FR Germany, 1975):

Category 1. *Respiratory toxicity*: expressed as LCt_{50} (mg min/m^3) for a minute volume of 20 litres of air. The chance that a particular dose will be received by a person when the air around him contains a toxic substance will be proportional to its concentration in the air and to the time for which he is exposed to this concentration. This function of concentration and time is referred to as the Ct. Generally, when the concentration changes with time, the total dosage received will be proportional to the area under the concentration–time curve, and the Ct will be equal to this area. Where the concentration is constant, the Ct is the product of the concentration and the time of exposure.

Category 2. *Percutaneous toxicity*: expressed as LD_{50} (mg/kg).

Category 3. *Skin lesion*: may be expressed as the dose in milligrams of substance per square centimetre (mg/cm^2) necessary to produce various skin changes (erythema, superficial blisters, deep blisters or necrotic ulceration).

A number of substances will produce toxic effects of more than one category. Mustard gas, for example, which causes severe skin lesions, is also highly toxic if inhaled, and VX is both a respiratory and a percutaneous poison. Besides, depending on the manner and duration of contamination, toxicity may take any of the following forms: *acute* toxicity, where a single exposure causes poisoning; *subacute* (or *subchronic*) toxicity, where repeated exposure over a relatively short period of time causes poisoning or puts the subject at risk; *chronic* toxicity, where repeated exposure to small doses over a long period of time causes poisoning or puts the subject at risk; or *delayed* toxicity, where single or repeated exposure to chemicals leads to toxic effects after a prolonged period of time (months or years) (SIPRI, 1975*b*: 1).

Most CW agents are liable to cause delayed effects. This applies not only to persons who have survived acute or subacute poisoning, but also to persons showing no signs of poisoning. These delayed effects may take different forms—psychopathological–neurological changes, malignant tumours, pathological changes in blood, mutagenic, teratogenic and embryotoxic effects, and so on. The same is true of a number of intermediates and possibly also of binary components.

No effective prophylactic or therapeutic measures are available at present against these delayed effects. This means that the handling of these poisons in the course of storage, transportation or destruction carries the risk of poisoning and can lead to delayed lesions (SIPRI, 1975*b*: 46).

Two or more chemicals may produce different toxic effects in the body:

1. They may have *independent* toxic effects, although these effects occur simultaneously.
2. They may *reinforce* each other's hazards.
3. Toxic hazards may be *additive*, chemicals with similar types of toxic action adding together their poisonous effects.
4. Some chemicals act *synergistically*, the presence of one substance greatly increasing the toxicity of the other.

Although acute poisoning is usually the most serious risk incurred during the storage or destruction of CW agents, additional hazards arise from the various chemicals used for detoxifying these agents. They include irritation and inflammation of the skin (dermatitis), eyes (conjunctivitis) and respiratory tract (pharyngitis, pneumonitis). Allergic responses often occur in repeated exposures to low concentrations of different chemical substances.

Tolerance level

Exposure limits (tolerance levels) are expressed for *gases* and *vapours* in parts per million (ppm) or in milligrams per cubic metre (mg/m^3). Both units are valid in most European countries for 20°C and 760 torr (760 mm Hg pressure), and in the USA and other countries for 25°C and 760 mm Hg pressure. The formulas used for calculation are:

$$\left.\begin{aligned} mg/m^3 &= \frac{\text{Mol. weight}}{24.04} \times ppm \\ ppm &= \frac{24.04}{\text{Mol. weight}} \times mg/m^3 \end{aligned}\right\} \text{ for } 20^\circ C$$

$$\left.\begin{aligned} mg/m^3 &= \frac{\text{Mol. weight}}{24.44} \times ppm \\ ppm &= \frac{24.44}{\text{Mol. weight}} \times mg/m^3 \end{aligned}\right\} \text{ for } 25^\circ C$$

The expression "exposure limits" appears in the Working Environment (Air Pollution, Noise and Vibration) Convention adopted by the International Labour Conference in 1977 (Preisich, 1978). As used in this publication, it is not intended to define the meaning of the values specified in the various national lists. It is used as a general term and therefore covers the various expressions employed in the national lists, such as "maximum allowable concentration" (MAC), "threshold limit value" (TLV), "permissible level", "limit value", "average limit value", "permissible limit", "time-weighted average", "industrial hygiene standard", and so on.

The criteria and methods for determining exposure limits are not the same in different countries. They vary in practice between the stringent Soviet concept of MAC, which in no case should produce biological or functional changes, and the more flexible approach of the American Conference of Governmental Industrial Hygienists (ACGIH) of the USA, whose TLV makes allowance for reversible clinical changes. These values are generally established on the assumption of an eight-hour shift exposure, work of normal intensity, normal climatic conditions and an exposure-free period of 16 hours following the shift, during which full detoxification should ensue.

The setting of internationally agreed tolerances is extremely important for CW-agent destruction in the framework of an international chemical disarmament treaty. Without such international agreement, there is always the danger that one country will set different tolerance levels than another.

For the toxicologist, it is important that MAC or TLV do not necessarily relate directly to systemic toxicity but may also be set on the basis of eye or skin irritation, narcosis, nuisance, or other forms of clinical manifestation that would be undesirable in the work environment. Thus, it would be dangerous to attempt to utilize TLVs as a basis for comparison of the systemic toxicity of chemicals without a thorough understanding of the information incorporated in the setting of the individual guide-lines.

By way of illustration, the estimated acute toxicity values, as TLVs, for some CW agents and other chemicals are given in tables 1 and 2 respectively.

Table 1. Estimated acute toxicity values of some chemical warfare agents

CW agent	LCt_{50} (mg min/m^3)	Threshold limit value (mg/m^3)
Lethal		
VX	10–50	0.00001
Sarin	100	0.0001
Soman	50–100	–
Tabun	400	0.003
Mustard	1 500	–
Lewisite	1 500	–
Phosgene	3 200	0.2
Hydrogen cyanide	5 000	11.0 (skin)
Cyanogen chloride	11 000	–
Adamsite	15 000	–
Chlorine	19 000	1.5
	ICt_{50} (mg min/m^3)	
Irritating		
CS	10–20	0.4 (skin)
CN	80	0.3
Arsine derivatives	10–30	–
Chloropicrin	–	0.7
	ICt_{50} (mg min/m^3)	
Incapacitating		
BZ	200	

Table 2. Estimated acute toxicity values of some chemicals important in destruction procedures for chemical warfare agents

Chemical	Estimated fatal dose (g/man)	Threshold limit value (mg/m^3)
Phosphorus trichloride	1	3.0
Isopropyl alcohol	250 ml	980.0
Ammonia	–	35.0
Chlorine	–	1.5
Fluorine	–	0.2
Hydrogen fluoride	–	2.0
Hydrogen chloride	1	7.0
Ethylene oxide	4 000 (ppm)	90.0
Sulphur dioxide	–	13.0
Sodium hydroxide	5	2.0
Carbon monoxide	12 010 (ppm)	55.0

III. Health control

It is the task of the medical services to take care of people handling CW agents and other chemicals in places of storage, during transport and at the destruction site. In view of the different toxicities of the various CW agents and other chemicals, appropriate medical and security measures will naturally need to be adopted.

The problem of panic in an emergency during the destruction of CW agents should never be underestimated, even in the case of highly trained personnel. This is particularly important in nerve-gas poisoning because of the direct toxicological effects of these gases on the central nervous system.

Periodic health control

It is necessary that a medical and occupational history of every prospective employee be taken and that he be given a physical examination before entering employment. Workers handling these hazardous and highly toxic materials should be in excellent physical and mental condition. Every worker coming in contact with CW agents and the various chemicals used in the destruction process—including the storage and transportation phases—should be under medical control throughout the duration of the work and should have an up-to-date health record incorporating all the necessary data. Workers engaged in the storage or destruction of nerve agents should be placed under special medical supervision. Pre-employment and subsequent periodic medical examinations should be conducted by a physician who is thoroughly conversant with CW-agent toxicology and its clinical manifestations. To begin with, it is prudent to exclude from the destruction operations all juveniles and aged persons, expectant and nursing mothers, persons suffering from psychiatric disorders, alcoholics, drug addicts and those otherwise debilitated by organic illness.

Physicians and other medical officers must be alert to the physical changes pathognomic of CW-agent poisoning. In addition, they should be on the look-out for more general reactions—such as those of the skin and the respiratory system and of the workers' general adaptability to the job.

Health education of personnel

In order to be able to use any tool—whether mechanical or chemical—safely, the user must be aware of the hazards involved and must also be familiar with the procedures and equipment necessary for avoiding such hazards. The most serious hazard to be avoided in work with CW agents is that of acute poisoning. Persons handling CW agents and other chemicals should know how these agents penetrate into the body; how to handle these agents; what kind of adverse effect may be expected; what the typical symptoms of poisoning are; and what to do in case of poisoning. The superintendent physician should satisfy himself that this basic information is well understood by all workers before they are assigned to work with chemicals, especially with CW agents.

Medical protection

The medical protection of people involved in all the various stages and operations of the destruction of CW agents entails organizational problems, staff problems and the problems of securing the necessary supplies.

The medical supervision programme, conducted by a physician, should include (*a*) a ready worked-out plan for adequate first aid and special care in emergencies, and (*b*) initial and periodic laboratory tests, medical examinations and other pertinent observations of personnel for as early detection as possible of any harmful effects of chemicals.

One of the functions of medical supervision is to monitor the safety programme. Any overexposure to toxic materials that occurs should be regarded as a failure in operation of the safety programme, demanding immediate attention and correction. The supervisory physician should therefore be well acquainted with the work of the personnel he supervises, and also with the CW agents and other chemicals they use.

Medical supervision is particularly valuable for workers handling the highly toxic organophosphorus CW agents. In this kind of poisoning, prompt and adequate first aid and medical attention can make the difference between life and death. The first aid and treatment of other types of CW-agent poisoning is rarely as specific and dramatic as in the case of nerve-gas poisoning.

Since nerve-gas poisonings are the most likely to occur, the following facts need to be emphasized:

1. As for the question of whether workers themselves should carry atropine or oxime for use in first aid, the answer is that they should not carry these antidotes.

2. If a medical supervision programme exists and if the attending physician can be absolutely certain that workers will be brought to him immediately on the slightest suspicion of poisoning, atropine or oxime should never be taken as preventive measures.

3. Oral administration of atropine or oximes has no place in a real poisoning emergency. The dose of atropine is too small to suppress poisoning; besides, the victim is incapable of oral medication when he is vomiting or stuporous. Moreover, because of the pharmacodynamic effects of atropine, increased doses of this drug would modify the working ability of personnel and could be dangerous for them (Vojvodić, Rosić, and Vojvodić, 1967). Increased doses of oximes would not save workers in cases of serious nerve-agent poisoning, especially if soman were involved (Vojvodić and Miletić, 1968). As in the case of oral administration of atropine, oximes can impart a false sense of security, delaying or preventing first-aid measures and prompt medical treatment (Vojvodić, 1973).

Ambulance crews, medical personnel, firemen, policemen and other officials may also be exposed while dealing with emergencies involving CW-agent contamination.

Health measures recommended by CAMDS[1]

The US Army has performed a great many experiments on the biology, chemistry and toxicology of CW agents as well as on protection, decontamination and disposal activities. Brief excerpts on health and medical problems encountered in the "demilitarization" procedures at the CAMDS destruction facility at Tooele Army Depot, Utah, USA appear in NTIS (National Technical Information Service, US Dept of Commerce) Bibliographies with Abstracts (NTIS, 1977, 1978) and a complete elaboration of destruction problems is presented in the Final Environmental Impact Statement on CAMDS (US Dept of the Army, 1977). Some points are cited below:

> Personnel involved in the actual operations will be given intensive training in handling chemical agents as well as in the operation and the maintenance of the demilitarization equipment to further insure system safety. Training will also cover plant safety requirements and safety regulations.

[1] CAMDS = Chemical Agent Munitions Disposal System (Tooele Army Depot, Tooele, Utah, USA).

If emission limits have been exceeded the protective measures for the plant workers should be done until corrective actions are taken.

A first aid kit properly marked and located will be maintained at all needed places with all the necessary provisions.

A site medical facility provides emergency medical treatment for the CAMDS staff. It consists of a decontamination and shower area, an emergency room and stand by quarters.

A casualty is brought into the decontamination and shower room where he is monitored. Clothing removed and casualty showered, monitored and taken in emergency room for treatment. After treatment the patient will be moved to a hospital by ambulance at the medical facility.

Medical support for CAMDS operations will be provided through an outside medical facility, staged 24 hrs per day by medical technicians, backed up by a secondary aid station at the South Area administrative area and medical doctors in the Tooele Army Depot North Area...

IV. Conclusions

In the Final Environmental Impact Statement referred to above (US Dept of the Army, 1977), it is said that extremely small quantities of CW agents will be released into the atmosphere during demilitarization processes, and that these emissions will not be allowed to exceed the limits established by the US Surgeon General as safe for operation. Apart from the lack of many details, nothing is said in the statement about the possible delayed effects on humans.

The statement does not adequately address the potential hazard that may exist to people living in the vicinity of the proposed detoxification facility. What is clearly needed is a description of the potential hazard in case hazardous substances should be released owing to failure of plant operations, and a discussion of the measures necessary for protecting the health of the public in such an event.

Both the points mentioned above—namely, the delayed effects and the potential hazard to people living in the vicinity of the detoxification facility—are of the utmost importance for any destruction plant to be constructed.

In connection with work on the destruction of CW agents and chemical munitions, special attention should be paid to the following points:

1. There should be regular health control of the personnel engaged in the destruction work as well as of all persons entering the storage area, patrolling guards, and so on.
2. There should be special treatment and permanent protection for personnel who accidentally come in contact with CW agents or other toxic chemicals.
3. Non-medical personnel should be trained to treat casualties.
4. Drugs for counteracting the toxic effects of CW agents and other chemicals should be stockpiled.

5. Additional hospital beds should be prepared for coping with a sudden influx of casualties.

Needless to say, all the above measures presuppose perfect functioning of the technical protection routines—monitoring, detection, decontamination, protective devices, and so on—because without them they are of no value.

References

Belcher, D. W., 1977. Spray drying of war gas residue. *CEP July* (reprinted by Stork Bowen Engineering Inc., Somerville, New Jersey, USA).

Canada, 1974. Disarmament Conference document CCD/434, 16 July.

FR Germany, 1975. Disarmament Conference document CCD/458, 22 July.

NTIS, 1977. *Chemical and Biological Warfare. Part 2. Protection, Decontamination and Disposal.* NTIS/PS-77/1030. Springfield, Virginia: National Technical Information Service, US Dept of Commerce.

NTIS, 1978. *Chemical and Biological Warfare. Part 3. Biology, Chemistry and Toxicology.* NTIS/PS-78/1297. Springfield, Virginia: National Technical Information Service, US Dept of Commerce.

Preisich, M. A., 1978. Survey on exposure limits of noxious substances. *FECS/WPPA/47, Budapest.*

SIPRI, 1971. Some thoughts on the problem of national and international control regarding organophosphorus compounds. In *Possible Techniques for Inspection of Production of Organophosphorus Compounds*, SIPRI Symposium Report, p. 19. Stockholm: Stockholm International Peace Research Institute.

SIPRI, 1975*a*. Methods of destruction of chemical warfare agents. In *Chemical Disarmament: New Weapons for Old*, pp. 102–13. Stockholm: Almqvist & Wiksell.

SIPRI, 1975*b*. *Delayed Toxic Effects of Chemical Warfare Agents*. Stockholm: Almqvist & Wiksell.

SIPRI, 1978. The destruction of chemical warfare agents. In *World Armaments and Disarmament: SIPRI Yearbook 1978*, pp. 360–76. London: Taylor & Francis.

SIPRI, 1979. Stockpiles of chemical weapons and their destruction. In *World Armaments and Disarmament: SIPRI Yearbook 1979*, pp. 470–89. London: Taylor & Francis.

USA, 1972. Disarmament Conference document CCD/367, 20 June.

USA, 1974*a*. Disarmament Conference document CCD/435, 16 July.

USA, 1974*b*. Disarmament Conference document CCD/436, 16 July.

USA, 1974*c*. Disarmament Conference document CCD/437, 16 July.

US Dept of the Army, 1977. *Operation of the Chemical Agent Munitions Disposal System (CAMDS) at Tooele Army Depot, Utah.* Final Environmental Impact Statement, Office of the Project Manager for Chemical Demilitarization and Installation Restoration, March.

Vojvodić, V., 1973. Clinical picture, diagnosis and treatment of poisoning by organophosphate insecticides. *Arh. hig. rada*, **24**: 341–55.

Vojvodić, V., and Miletić, S., 1968. The effect of toxogonin in men poisoned by non-lethal doses of soman. In *First Yugoslav Symposium on Medical Toxicology, Selected Topics*, pp. 287–93. Belgrade (in Serbo-Croatian, English summary).

Vojvodić, V., Rosić, N., and Vojvodić, M., 1967. Effects of atropine sulphate on the body and some elements of fighting capability of healthy volunteers. *Vojnosanit. Pregl.*, **24**: 10 and 522–26.
Yugoslavia, 1972. Disarmament Conference document CCD/377, 20 July.
Yugoslavia, 1976. Disarmament Conference document CCD/503, 5 July.
Yugoslavia, 1976. Disarmament Conference document CCD/504, 5 July.

Biomedical aspects of the destruction and conversion of chemical warfare agents

L. ROSIVAL
Centre of Hygiene of the Research Institute for Preventive Medicine, Bratislava, Czechoslovakia

Abstract. The WHO global medium-term programme for the promotion of environmental health is an integral part of world-wide endeavours for attaining safety from chemical hazards.

The terms *toxicity*, *hazard* and *risk* are defined. Occupational and environmental hazards are discussed. The functions of environmental and health monitoring are evaluated with respect to occupational exposure. Environmental-health problems associated with the destruction and conversion of chemical warfare agents—for example, air and water quality, wastes, and ecological and health aspects—are discussed.

I. The WHO programme for the promotion of environmental health

In his 1978 biennial report, the Director-General of the World Health Organization announced the WHO global medium-term programme for the promotion of environmental health (WHO, 1978*a*). The report stresses the priorities and needs of member states. One of the four component programmes is recognition and control of environmental conditions and hazards affecting human health. This component programme includes activities for evaluating the effects of chemicals on human health, the dissemination of information about these activities with a view to attaining a rational basis for the strategy and tactic for planning national control programmes, the exchange of information on measures for the technical prevention of adverse effects, and preparations for manpower training. One of the most important aims is to establish and support international programmes for chemical safety (criteria, standards, etc.). In this connection, extensive evaluations of several chemicals have been made in recent years, six criteria documents have been published and the health aspects of some industries have been reviewed. The health effects have also been considered from the methodological angle, and toxicological and epidemiological methods have been evaluated.

This programme is a good example of co-operation between WHO and member states in developing environmental-health projects.

Toxicity, hazard and risk

The problem of the destruction and conversion of chemical weapons and bulk chemical-warfare (CW) agents may be regarded as an integral part of the above-mentioned global movement for safety from chemical hazards.

From the biomedical point of view there are several aspects that must be taken into account in the evaluation of the occupational and non-occupational exposure of human beings. In this connection it is important, first of all, to define the terms *toxicity*, *hazard* and *risk* (WHO, 1978 *b*). The *toxicity* of a substance is its capacity to cause injury to a living organism. On the basis of toxicity, CW agents fall into two groups: one group consists of supertoxic compounds, the other comprises those substances that give rise to toxic effects in high concentration. *Hazard* indicates the likelihood that a chemical will cause an adverse health effect under the conditions in which it is produced or used. *Risk* is the expected frequency of undesirable effects arising from exposure to a chemical.

II. Occupational exposure

As regards occupational exposure, there are some important problems relating to environmental and health monitoring. Monitoring has two functions: first, the making of routine measurements on health and environmental indices and the recording and transmission of these data; and, second, the collation and interpretation of such data with a view to detecting changes in the health status of populations and in their environment (WHO, 1973). There are two types of monitoring: environmental monitoring and medical monitoring.

Environmental monitoring

In view of the high toxicity and rapid physiological action of CW agents, simple and very sensitive methods for their detection in the environment are necessary. The existing colorimetric methods are not sufficiently sensitive (sensitivity: *c.* 1 μg for sarin and tabun, and 2 μg for DFP). These methods are based on the acceleration of oxidation of an amine base (benzidine) in the presence of G-agents. Methods of this type are selective for organophosphorus compounds in general. The sensitivity and time factor are critical for all such chemical detection methods.

Methods based on fluorescence appear to be suitable for the detection and determination of G-agents. The Schoenemann reaction is well known. It is based on the oxidation of *o*-tolidine with peroxide in alkaline medium in the presence of G-agents. This reaction has been modified for the quantitative evaluation of G-agents. The method is sufficiently specific and the sensitivity is 100 to 1 000 times greater than that of colorimetric methods. Experimental studies have shown that indol can be converted to indigo by using alkaline peroxide in the presence of G-agents. The sensitivity for G-agents is of the order of 1 μg. The critical point seems to be the short-lived fluorescent state of the product concerned.

A portable alarm for the detection of toxic agents in the atmosphere was developed on the basis of the Schoenemann reaction. Some parameters for this instrument are: a response time of 2 minutes for a concentration of GB (sarin) of 0.22 ppm, or 5 seconds for a concentration of 2 ppm. The speed of air-flow through the instrument is 1.5 litres per minute. A variant of this instrument—a portable analyser for nerve gases—was also developed on the basis of the Schoenemann reaction, using the conversion of indol to fluorescent indoxyl. This instrument may respond to less than 1 μg of GB per litre of air.

In recent years there has been a dynamic development of sophisticated analytical methods that enable the detection of chemicals to be performed with sufficient speed and reliability—for instance, gas-liquid chromatography in combination with spectroscopic methods.

Medical monitoring

This type of monitoring makes it possible to identify persons likely to be vulnerable to certain exposures and provides baseline data for measuring early adverse effects of exposure in persons at risk.

For unprotected workers it is necessary to fix the maximum allowable concentrations for GB, VX, mustard and other chemicals for periods of one, three and more hours in each case.

Apart from ensuring that the various statutory requirements are being observed in "demilitarization" plants, the medical staff will be involved in the following tasks: preparatory assessments, periodic and special medical assessments, data collection, information and instruction, preparation of an emergency plan, and interpretation of toxicity, hazard, risk and safety data. Management and workers must be kept fully aware of all the dangers and must employ optimum safety measures. It is a *sine qua non* that the workers are kept fully informed about all the hazards associated with the chemicals they are handling.

III. Environmental protection

The operation of demilitarization plants creates complex problems and may have a serious impact on the environment. Among the factors that have to be considered are air and water quality, wastes, and the ecological and human-health aspects. Any proposed measures for solving these problems must take account of the ecotoxicological aspects. These include such factors as the emission, entry, distribution and fate of chemicals in the abiotic environment. The next step is to study the entry and fate of chemicals in the biosphere, not overlooking the possibility of contamination of biological chains. The last link in this ecochemical chain is a study of the toxic effects of chemicals on ecosystems and on man.

For the operation of demilitarization plants it is suggested that the problems of CW agents be dealt with analogously to those of radioactive wastes from nuclear reactors. Thorough studies should be made of the best disposal methods that entail no health hazards to the population and no environmental pollution. Lethal chemical agents should not be dumped in the ocean; the work of destruction should be carried out on land. Many specific technological problems (not taken up in this paper) are involved, and there should be a step-by-step evaluation of the different systems used (prototypes of mobile demilitarization systems, retention in storage, methods for final disposal, etc.) and an exchange of information at international level.

For each system, destruction operations must be preceded by a clear description of the intended action (objectives, technological destruction steps, and specification of lethal agents and inert materials to be destroyed or deposited). In some cases very large amounts of agents, explosives and neutralizing agents (sodium hydroxide, chlorine, etc.) are involved and the chemical nature of the waste must be specified. These precautions are necessary in view of the potential hazard to populations living in the vicinity of the destruction facilities.

Disposal facilities and equipment must be clearly described, and it is important to make provision for medical facilities throughout the operations entailed. Information must be supplied on the exact location of storage facilities because of the geological hazards involved in high-risk seismic regions.

Every precaution should be taken to prevent the release of toxic substances into the environment throughout the demilitarization process. Special monitoring of the air and water is of the utmost importance from the toxicological point of view. In several countries there are fixed standards for pollutants in the ambient air, such as suspended particulates, sulphur dioxide and oxides of nitrogen. These standards sometimes vary within a broad range, but they nevertheless form a good basis for establishing technical and medical preventive measures. In some countries standards have also been prescribed for HD (mustard gas), GB and VX that can be very helpful in implementing preventive measures.

The problems presented by solid wastes must be solved satisfactorily. Criteria have been worked out in several countries for the disposal of solid toxic chemical substances, and international coordination is necessary. Disposal methods need to be studied carefully because the wastes can vary greatly in structure.

As regards ecotoxicological problems, it is important to promote research in the field of ecotoxicology with the aim of ensuring that no step in the entire demilitarization process has an adverse effect on the various ecosystems.

The retention of an assortment of toxic chemical agents in storage is not to be recommended (danger of leakage, high monitoring costs, etc.). The aim must be to develop technology in line with the criteria and standards laid down by the national control authorities.

Recent years have witnessed successful international collaboration in industry between health authorities on the one hand, and chemists, hygienists, experimental and clinical toxicologists, safety experts, biologists, ecologists, etc., on the other. This collaboration has had the support of both management and labour.

We are now living in a time of a "toxicological wave" that confronts us with enormous new dramatic challenges. From the biomedical standpoint, much remains to be done also in the field of the destruction and conversion of lethal chemical agents as an integral part of comprehensive safety from chemical hazards.

References

WHO, 1973, Wld Hlth Org. techn. Rep. Ser., No. 535.

WHO, 1978 *a*. Promotion of environmental health. In *Official Records of the World Health Organization No. 243*, The Work of WHO 1976–1977, pp. 151–62. Biennial report of the Director-General to the World Health Assembly and to the United Nations. Geneva: World Health Organization.

WHO, 1978 *b*. *Environmental Health Criteria 6: Principles and Methods for Calculating the Toxicity of Chemicals, Part I*. Geneva: World Health Organization.

Long-term effects of acute exposure to nerve gases upon human health

B. BOŠKOVIĆ and R. KUŠIĆ
Military Technical Institute, Medical Department, Belgrade, Yugoslavia

Abstract. For many years it has been maintained that acute poisoning by organophosphorus insecticides and nerve gases causes only functional but not histopathological changes. However, recent detailed studies of acute poisoning by organophosphorus anticholinesterases have revealed a high incidence of paralysis in humans which differed from the signs of "delayed neurotoxicity", as well as skeletal muscle necrosis in rats. Furthermore, the latest results in monkeys and humans show that acute exposure to toxic doses of sarin produces prolonged changes in the brain function. These results point to the long-term risks to the health of populations exposed to nerve gases and call for an immediate ban on chemical warfare agents and on their study in voluntary human subjects.

The extensive use of organophosphorus compounds (OPC) in the past three decades has led to numerous cases of acute and chronic poisoning by these compounds. Consequently, considerable experience in the therapy of poisoning by these compounds has meanwhile been gained.

According to previous findings in humans (Grob *et al*., 1950) and in experimental animals (Holmstedt *et al*., 1957), it was generally held that acute poisoning by OPC is followed only by functional but not by histopathological changes. The principal pathological findings—usually not striking—were capillary dilatation, hyperaemia and oedema, occurring most commonly in the lungs, but appearing also in brain, liver, spleen, kidneys and subcutaneous tissue. Some recent results of the liver and kidney damage in acute experimental poisoning by parathion (*O,O*-diethyl *O-p*-nitrophenyl phosphorothioate) (Gaaz *et al*., 1974), paraoxon (*O,O*-diethyl *O-p*-nitrophenyl phosphate) and soman (pinacolyl methylphosphonofluoridate) (Domschke and Domagk, 1970; Domschke *et al*., 1971; Hettwer, 1975*a*, 1975*b*) were ascribed to the non-specific nature of intoxication—namely, to intracellular hypoxia—and spoke in favour of such an assumption.

On the other hand, the toxic effects followed by pathological changes after prolonged occupational exposure to nerve gases and OP insecticides are today clearly established and have been summarized in a monograph published by SIPRI (1975).

However, recent growing data show that acute poisoning by acetylcholinesterase (AChE) inhibitors may also cause ultrastructural morphological changes. Thus, for example, Laskowski *et al*. (1977) and Wecker *et al*. (1978) have found that paraoxon produces myopathic changes in rats, characterized by dilated mitochondria, expanded sarcoplasmic reticulum, accumulation of coated vesicles in nerve terminals and widening of subsynaptic folds often containing membrane-bound vesicles. Maximal skeletal muscle fibre necrosis in their experiments depended on a critical degree and duration of AChE inhibition, while reactivation of AChE by 2-PAM Cl (2-pyridine aldoxime methyl chloride) prevented the myopathic changes. It is also interesting to note that soman exerts marked "direct" toxic biochemical effects outside the acetylcholine–acetylcholinesterase (ACh–AChE) system (Bošković, 1979).

In cases of suicidal ingestion of OP insecticides, Wadia *et al.* (1974) found paralysis which differed from the signs of "delayed neurotoxicity", the mechanism of which was elucidated by Aldridge and Johnson (1971). Wadia *et al.* (1974) attributed the high incidence of paralysis (26% of cases) to the rare and inadequate use of 2-PAM Cl in the treatment of their patients.

It has also been shown that poisoning by a single administration of sarin (isopropyl methylphosphonofluoridate) alters the electroencephalogram (EEG) in Rhesus monkeys for up to one year (Burchfiel *et al.*, 1976). Furthermore, significant changes in waking and sleeping EEGs have been found in industrial workers with histories of even a single accidental exposure to toxic levels of sarin (Duffy *et al.*, 1979). These results point to the long-term changes in the brain function of primates exposed to acute sarin poisoning, at a time when tissue AChE activity was probably within normal values.

Duffy *et al.* (1979) have offered a few explanations for these findings. First, the low tissue AChE activity and ACh excess at the time of exposure can induce long-term changes in synaptic morphology or biochemical organization which do not completely reverse when tissue enzyme activity returns to normal. Second, high ACh concentrations at the time of exposure might produce long-term changes in the postsynaptic receptor which render it more sensitive to endogenous ACh. Third (but the least probable), it is possible that sarin has actions beside that of a potent anticholinesterase and that, similarly to TOCP (tri-*o*-cresyl phosphate) and DFP (diisopropyl fluorophosphate), it produces neuropathies of delayed onset. Since sarin is not reported to produce delayed neurotoxic responses, its persistent central-nervous-system (CNS) effects in primates cannot be related to a direct action on central axons or myelin. The authors conclude that "regardless of the pathogenic mechanisms, results of the current study confirm the ability ... that OP exposure can produce long-term change in the brain function of ... humans."

These results point to the marked hazards arising from exposure to toxic levels of nerve gases. Moreover, owing to the inadequacies of the existing methods used in routine control of human health, one cannot rule out some other, so far not detectable, changes in these poisonings.

Nowadays, the groups of people most dangerously exposed to the toxic effects of nerve gases are those engaged in work on their production, storage, handling, and destruction. However, in the case of warfare use of nerve gases, the proportion of possible long-term consequences in people who survive an acute poisoning may represent the unforeseeable hazards. This is an important additional argument for an immediate and complete ban on the production and use of chemical warfare agents.

For the above reasons, we strongly support the proposition by SIPRI (1975) on the prohibition of experiments with nerve gases on volunteers. In one such experiment a large number of people were recently exposed to toxic doses of VX (*O*-ethyl *S*-2-diisopropylaminoethyl methylphosphonothiolate) (Craig *et al.*, 1977).

References

Aldridge, W. N., and Johnson, M. K., 1971. Side effects of organophosphorus compounds: delayed neurotoxicity. *Bull. W.H.O.*, **44**:259–263.

Bošković, B., 1979. The influence of 2-(*o*-cresyl)-4 H-1 : 3 : 2-benzodioxa-phosphorin-2-oxide (CBDP) on organophosphate poisoning and its therapy. *Arch. Toxicol.*, **42**:207–216.

Burchfiel, J. L., Duffy, F. H., and Sim, V. M., 1976. Persistent effects of sarin and dieldrin upon the primate electroencephalogram. *Toxicol. Appl. Pharmacol.*, **35**:365–379.

Craig, N. F., Cummings, G. E., and Sim, V. M., 1977. Environmental temperature and the percutaneous absorption of a cholinesterase inhibitor, VX. *J. Investig. Dermatol.*, **68**: 357–361.

Domschke, W., and Domagk, G. F., 1970. Enzyminduktion in der Rattenleber durch Soman. *Naturwissenschaften*, **57**:39.

Domschke, W., Domschke, W., and Classen, M., 1971. Zum Mechanismus der Leberzellschädigung durch Alkylphosphate. *Naturwissenschaften*, **58**:575.

Duffy, H. F., Burchfiel, J. L., Bartels, H. P., Goan, M., and Sim, V. M., 1979. Long-term effects of an organophosphate upon the human electroencephalogram. *Toxicol. Appl. Pharmacol.*, **47**:161–176.

Gaaz, J. W., Poser, W., and Erdmann, W. D., 1974. Untersuchungen zur Lebertoxicität von Nitrostigmin (Parathion, E-605) und der perfundierten Ratten-leber. *Arch. Toxicol.*, **33**:31–40.

Grob, D., Garlick, W. L., and Harvey, A. M. G., 1950. The toxic effects in man of the anticholinesterase insecticide parathion (*p*-nitrophenyl diethyl thionophosphate). *Bull. Johns Hopkins Hosp.*, **87**:106–129.

Hettwer, H., 1975*a*. Über Veränderungen an der Ultrastruktur der Rattenniere nach Phosphosäureestervergiftung. *Acta histochem.*, **52**:165–210.

Hettwer, H., 1975*b*. Histochemische Untersuchungen an Leber und Niere der Ratte nach Phosphosäureestervergiftung. *Acta histochem.*, **52**:239–252.

Holmstedt, B., Krook, L., and Rooney, J. R., 1957. The pathology of experimental cholinesterase-inhibitor poisoning. *Acta Pharmac. Toxicol.*, **13**:337–344.

Laskowski, M. B., Olson, H. W., and Dettbarn, W. D., 1977. Initial ultrastructure abnormalities at the motor end plate produced by a cholinesterase inhibitor. *Exp. Neurol.*, **57**:13–33.

SIPRI, 1975. *Delayed Toxic Effects of Chemical Warfare Agents*. Stockholm: Almqvist & Wiksell.

Wadia, S. R., Sadagopan, C., Amin, B. R., and Sardesai, V. H., 1974. Neurological manifestations of organophosphorus insecticide poisoning. *J. Neurol. Neurosurg. Psych.*, **37**:841–847.

Wecker, L., Kiauta, T., and Dettbarn, W. D., 1978. Relationship between acetylcholinesterase inhibition and the development of a myopathy. *J. Pharmac. Exp. Ther.*, **206**:97–104.

Some aspects of the problem of the destruction of chemical warfare agents

O. A. REUTOV
Department of Chemistry, University of Moscow, Moscow, USSR

and K. K. BABIEVSKY
Institute of Organoelement Compounds, USSR Academy of Sciences, Moscow, USSR

Abstract. While international negotiations on banning chemical weapons have been going on for many years, no consensus has been reached on certain important aspects of this problem—for example, the destruction of chemical weapons.

It is generally felt that on-site inspection of the destruction of chemical weapons raises apprehensions concerning both trade and military secrets.

The destruction of stockpiles of chemical warfare (CW) agents should be controlled by representatives of national control agencies acting in co-operation with an international consultative committee. The functioning of such possible agencies and their interaction with the proposed international consultative committee are matters that need further study.

I. The destruction of chemical warfare agents

While international negotiations on banning chemical weapons have been going on for many years, no consensus has been reached on certain important aspects of this problem—for example, the destruction of chemical weapons.

The question of the destruction of chemical weapons was first raised in the disarmament context more than 50 years ago. At a meeting of the Preparatory Commission for the Disarmament Conference held on 23 March 1928, the Soviet delegate put forward a proposal for a protocol, intended to supplement the Geneva Protocol of 1925, that included the following provision:

> *Article 1.* All methods of and appliances for chemical aggression (all asphyxiating gases used for warlike purposes, as well as all appliances for their discharge . . .) . . . whether in service with troops or in reserve or in process of manufacture, shall be destroyed within three months of entry into force of the present Convention. (League of Nations, 1928)

The destruction of stocks of chemical and biological (CB) weapons was also discussed in 1932 at the Disarmament Conference in the general debate on chemical and bacteriological weapons, proposals being put forward by Italy and Norway, among others (League of Nations, 1932*a*, 1932*b*).

On 19 April 1954, the United Nations Disarmament Commission decided to establish a subcommittee, with a mandate to discuss in detail various disarmament proposals, among them the problem of chemical weapons. Some of the documents worked out by the subcommittee specifically mentioned the destruction of chemical warfare (CW) agents.

In the "Declaration on General and Complete Disarmament" presented on 19 September 1959 to the UN General Assembly, the Soviet Prime Minister urged that all stockpiles of chemical and bacteriological weapons in the possession of states should be removed and destroyed.

In 1960 the USA, in its "Program for General and Complete Disarmament under Effective International Control", stated that in the third stage of general and complete disarmament, all armaments, including weapons of mass destruction and vehicles for their delivery, would be destroyed or converted to peaceful uses. In the "Joint Statement of Agreed Principles for Disarmament Negotiations" of 1961, the USA and the USSR recommended that the programme for general and complete disarmament should contain provisions for the elimination of all stockpiles of nuclear, chemical, bacteriological and other weapons of mass destruction and the elimination of all means of delivery of such weapons.

The following measures for the destruction of chemical weapons were put forward for general and complete disarmament by the USSR on 15 March 1962 at the ENDC: "All kinds of chemical, biological and radiological weapons, whether directly attached to the troops or stored in various depots and storage places, shall be eliminated from the arsenals of States and destroyed (neutralised)" (USSR, 1962 *a*, 1962 *b*).

In 1969 and 1970, the international debate on CB weapons was stimulated by greater public information on the nature and dangers of these weapons, including the publication of a report by the UN Secretary-General on "Chemical and Bacteriological (Biological) Weapons and the Effects of their Possible Use" as well as a report by the World Health Organization (WHO) on "Health Aspects of Chemical and Biological Weapons".

On 19 September 1969, the USSR and other Socialist countries submitted to the UN General Assembly a draft convention on CB weapons which contained (Article 2) prerequisite conditions for undertaking, within a specified period of time, the destruction or diversion to peaceful uses of all previously accumulated CB weapons (UN, 1969). In the revised draft convention of 23 October 1970, the obligation to destroy stockpiles of CB weapons or to divert them to peaceful uses was extended to include equipment and means of delivery for such weapons (UN, 1970).

The methods for destroying CB weapons were first discussed in 1971, when Sweden reviewed the possible technical means of disposal and destruction and submitted a working paper to the Conference of the Committee on Disarmament (CCD) (Sweden, 1971). In 1972 other technical working papers and a "Work Program Regarding Negotiations on Prohibition of Chemical Weapons" were tabled by the US delegation. Some of these documents discussed the methods for the destruction of chemical weapons (USA, 1972 *a*, 1972 *b*).

On 28 March 1972, the Soviet Union and other Socialist countries submitted a paper entitled "Draft Convention on the Prohibition of the Development, Production and Stockpiling of Chemical Weapons and on their Destruction" to the CCD (Bulgaria *et al.*, 1972). The text of this draft is similar in scope to that of the Biological Weapons Convention. Article II states:

Each State Party to this Convention undertakes to destroy, or to divert to peaceful purposes, as soon as possible but not later than .... months after the entry into force of the Convention, all chemical agents, weapons, equipment and means of delivery specified in Article I of the Convention which are in its possession or under its jurisdiction or control.

Another attempt at finding a constructive solution to the problem of chemical disarmament was made in a working paper presented to the CCD by a group of 10 non-aligned nations (Argentina *et al.*, 1973). Also in 1973, a new proposal was put forward by the Japanese delegation to the CCD which, like the draft convention submitted by the Socialist countries, set up as its ultimate goal the comprehensive prohibition of all chemical weapons, and it dealt with some aspects of the destruction of stockpiles of chemical weapons (Japan, 1973).

II. Verification of the destruction of stockpiles of CW agents

It is generally felt that on-site inspection of the destruction of chemical weapons raises apprehensions concerning civilian destruction processes and military secrets. Hence, it is important to establish the level of intrusion required for reasonable assurance that a chemical disarmament convention will be observed.

In our opinion, the destruction of stockpiles of CW agents should be controlled by representatives of the respective national control agency, co-operating effectively with an international consultative committee. The structure and functions of national control agencies were discussed in detail at a Working Group meeting convened by SIPRI in December 1972 (SIPRI, 1973: 19–20 and 36–50). The Pugwash Chemical Warfare Study Group (see footnote 2 on page 1) has taken the view that national control agencies would necessarily have heavy responsibility for ensuring treaty compliance. The respective national control agencies should verify the completeness of the destruction of CW-agent and -munition stockpiles and also verify that contamination of the environment has not occurred.

General methodological problems for the organization of a control on the destruction of CW agents have often been discussed. To give a good example, the interrelationships between various destruction activities, and between destruction and other activities to be covered by a chemical disarmament convention have been successfully summarized by the Pugwash CW Study Group by means of diagrams (Lundin, 1978).

Conversion of CW plants to the production of compounds for peaceful use should be carried out under the on-site supervision of representatives of the national control system. In this case it seems reasonable that monitoring devices should be installed so as to keep track of the various products whose manufacture is under control. The monitors should be sealed and be accessible only to representatives of the national control agency. It is obvious that for the control on production of dual-purpose materials—especially organophosphorus compounds—an approach mainly towards statistical methods should be developed.

Following the destruction of existing stockpiles of chemical munitions and/or bulk agents, an official declaration to that effect should be made by the governments of the countries that have signed a chemical disarmament convention.

One of the conclusions drawn at the Working Group meeting convened by SIPRI in December 1972 was: "There is need for an international organization to fulfil several functions which would make an economic-data reporting and monitoring system a useful part of an overall verification scheme" (SIPRI, 1973: 128–29). In our opinion, a consultative committee may serve as a model for such an international body as proposed by the Pugwash CW Study Group (Lundin, 1978).

The Pugwash CW Study Group did not, however, discuss in detail how the consultative committee might be constituted or how it might operate. It was only agreed that, depending on the manner in which these matters could be resolved, there might or might not be a need for an international verification agency as well (Pugwash, 1978).

It seems possible that the consultative committee would need to have a sample analysis laboratory for standardizing analytical and data-reporting methods.

We hope that the Pugwash CW Study Group will initiate further studies on (*a*) the structure of a sample analysis laboratory, (*b*) elaboration of a national control system, and (*c*) the role of national verification agencies and their interaction with the proposed international consultative committee.

References

Argentina, Brazil, Burma, Egypt, Ethiopia, Mexico, Morocco, Nigeria, Sweden and Yugoslavia, 1973. Disarmament Conference document CCD/400, 26 April.

Bulgaria, Czechoslovakia, Hungary, Mongolia, Poland, Romania and USSR, 1972. Disarmament Conference document CCD/361, 28 March.

Japan, 1973. Disarmament Conference document CCD/413, 21 August.

League of Nations, 1928. Annex 5 to the Minutes of the Fifth Session of the Preparatory Commission for the Disarmament Conference. Document C.P.D. 115.

League of Nations, 1932*a*, Series of Publications: 1932.IX.32; Conf. D.106.

League of Nations, 1932*b*. Conference for the Reduction and Limitation of Armaments; Conf. D/Bureau/14.

Lundin, S. J., 1978. "On the question of destruction of chemical weapons". Background paper for the 6th Pugwash Chemical Warfare Workshop, Salt Lake City/Kansas City, 8–12 May.

Pugwash, 1978. Report of the 6th Pugwash Chemical Warfare Workshop. *Pugwash Newsletter*, **16**(1), July 1978, p. 19.

SIPRI, 1973. *Chemical Disarmament: Some Problems of Verification*. Stockholm: Almqvist & Wiksell.

Sweden, 1971. Disarmament Conference document CCD/324, 30 March.

UN, 1969. United Nations document A/7655, 19 September.

UN, 1970. United Nations document A/8136, 23 October.

USA, 1972*a*. Disarmament Conference document CCD/360, 20 March.

USA, 1972*b*. Disarmament Conference document CCD/367, 20 June.

USSR, 1962*a*. Disarmament Conference document ENDC/2, 15 March.

USSR, 1962*b*. Disarmament Conference document ENDC/PV.36, 14 May.

Verification of the destruction of stockpiles of chemical weapons

A. J. J. OOMS[1]
Prins Maurits Laboratory TNO, Rijswijk, The Netherlands

Abstract. A treaty prohibiting the development, production and stockpiling of chemical weapons will most certainly contain a paragraph on the destruction of the existing stockpile. Two problems can be distinguished in the verification of this process. First, is it possible to verify that a declared stockpile has indeed been destroyed? The answer to this question is affirmative if on-site inspection is permitted. In this case, however, care has to be taken to safeguard any military and industrial secrets by carefully selecting the inspectors; fortunately, there are already quite a number of precedents in this field. If the type of agent to be destroyed is declared, it will not be necessary to establish the exact chemical structure of the compound. In the case of a completely new type of agent, the inspection can be confined to toxicity measurements; such a case will probably be rather the exception than the rule.

The second question concerns the actual size of the stockpile of a country. What does it mean for a country's capability to engage in chemical warfare when it declares that a certain quantity of chemical agents from its stockpile is to be destroyed? The answer to this question is much more difficult, but if some assumptions made about the method of stockpiling are correct, a reasonable assessment seems possible.

I. Introduction

In all discussions on a prohibition of the development, production and stockpiling of chemical weapons which have been held at the Conference on Disarmament (CD) and its preceding bodies (the Eighteen Nation Disarmament Committee, ENDC; the Conference of the Committee on Disarmament, CCD) there seems to be a consensus of opinion that an eventual treaty should include a provision for the destruction (or conversion to peaceful use) of stockpiled chemical agents and munitions. Indeed, all three draft treaties tabled so far contain such a paragraph. As with a great many other issues in the field of arms control and disarmament, the question of verification of such an operation immediately arises.

The following topics are discussed: first, the problem of the size and shape of the stockpiles in the countries preparing for the eventuality of chemical warfare (CW) and the case of the so-called hidden stockpiles; second, methods of destruction; third, ways in which these methods can be verified; and finally, some conclusions will be drawn.

[1] The views expressed in this paper are those of the author and do not necessarily reflect the views of the Netherlands government or those of the Netherlands Organization for Applied Scientific Research TNO.

II. Stockpiles of chemical agents and their verification

In order for a country to engage in chemical warfare of some intensity it has to have access to a stockpile of chemical agents. Part of this stockpile will probably consist of fairly large containers for the chemical agents from which munitions and other dispensers can be charged. Some of these containers will probably be quite large (several cubic metres), others will be of a more modest size. Typical examples of the latter category are the so-called ton containers described in the Final Environmental Impact Statement (US Dept of the Army, 1977), containing roughly 700 kg of agents, or the containers shown at the dismantled Chemical Defence Establishment's Process Chemical Division, Nancekuke (UK) to a group of experts on chemical warfare during a visit in March 1979 in the framework of the Conference on Disarmament.

Another part of the stockpile will consist of chemical agents filled in munitions, such as artillery shells, rocket warheads, aerial bombs and mines. The aforementioned Environmental Impact Statement gives a description of quite a large number of typical chemical munitions.

The actual size of such a stockpile is rather difficult to establish. Over the years many estimates have been published of the US and Soviet stockpile sizes, ranging from tens of thousands to hundreds of thousands of tonnes (SIPRI, 1973). One must realize, however, that release of this type of information at particular points in time may have other purposes than simply to provide information to the public by the intelligence communities.

In stockpiling chemical weapons two rather contradictory ways are probably used. For a chemical weapon to be used, either in first use or in retaliation, a supply of such munitions has to be at hand. That means that these munitions must be stockpiled in moderate amounts over a very wide range of sites—i.e., the supply has to be decentralized. On the other hand, chemical weapons are very dangerous objects and must accordingly be protected carefully against accidents and secured against theft, sabotage or unauthorized use. These measures can be carried out at a few fixed locations, under heavy guard and with numerous, observable security precautions, since it is obviously far easier and cheaper to maintain the physical security of weapons under such conditions. As peace-time is still the more "normal" state, it is probable that by far the largest part of any national stockpile is kept in a few centralized locations. These centres are probably well concealed and hardened in order not to invite a pre-emptive first strike. On the other hand, they may very well be part of larger military installations and arsenals, probably kept under close observation by countries having access to observation satellites.

It is probable that the intelligence organizations of the USA and the USSR are aware of stockpile sites of chemical weapons and have made reasonable guesses of their sizes.

If a country wishes particularly to conceal a part of its chemical weapon stockpile, it can be done, but such clandestine stockpiles would probably have to be limited in size. This means that if a country decides as a first step in the process of destruction to declare the size of the stockpile it wishes to destroy or have destroyed, the intelligence communities will be able to make a reasonable guess as to what that means *vis-à-vis* the total estimated size of the residual stockpile.

As previously stated, it will never be possible to make an accurate estimation of a nation's chemical weapons stockpile in the same way as estimates of other weapon systems are made. Besides, the stockpile is not static. If, however, the amount of chemical agent destroyed—as verified—is a major part of the size of the estimated stockpile, then this will be in itself a very definite confidence-building measure.

It follows that stockpile destruction must be accompanied by destruction or dismantling of the stockpiling facilities.

III. Destruction of stockpiles

The problem of the eventual destruction of chemical stockpiles must have existed for as long as nations have prepared for chemical warfare. In the course of the years several agents which were at one time considered for use in chemical warfare have become obsolete and have been replaced by other more modern agents. In other cases, stockpiles of agents must have deteriorated to an extent that has made them unsuitable for use in chemical warfare. In the case of stockpiles of chemical munitions, obsolescence and potential malfunctioning must have been recurring problems. In all such cases, the owners of the stockpiles must eventually have resorted to destruction.

Not very much is known about procedures used by the potential chemical warfare states in the past, but it would probably be safe to assume that the methods used were of a rather primitive nature. The easiest way to dispose of unwanted chemicals is probably to bury them in remote spots, preferably at military installations. There is no hard evidence indicating that this has been done, but it is probably safe to assume that this procedure has been used in the past. Illegal dumping of dangerous materials which did occur a few years ago—at least in Europe—points to the fact that this method of "destruction" has its attractions.

The best-known case of stockpile destruction was carried out in a kind of disarmament effort after the defeat of Germany in 1945, during which large quantities of chemical stockpiles (including nerve agents and mustard gas) were dumped by the Allies in the ocean. The experiences of Danish fishermen in the Baltic have particularly shown that this type of destruction was not of a permanent nature.

Another well-known example of dumping in the sea was Operation CHASE, in the course of which the United States sank a number of concrete-encased munitions filled with nerve agents in the Caribbean in 1970.

From the environmental standpoint these rather crude methods are no longer acceptable and are probably being replaced by more sophisticated procedures.

Since another paper in this series deals exclusively with the destruction methods (see the paper by Lohs on pages 67–75) suffice it to say that a number of working papers of the CCD (CCD/324, 360, 366, 367, 381, 403, 434, 436, 485, 497, 498, 506, 538 and 539) mention methods for the destruction of chemical agents, whereas the above-mentioned Final Environmental Impact Statement gives rather detailed information about the methods used in the United States. On the one hand, these methods consist in pyrolysis at rather high temperatures, with scrubbing of the released gases and vapours, and on the other, chemical reactions with alkaline or chlorinating reagents, and possibly catalytic decomposition.

The above-mentioned environmental problems will slow down the entire procedure of CW-agent destruction. In addition, there will be the problem of having to deal with large amounts of decomposition products, which would otherwise have been disseminated in the air or in river-water.

IV. Verification of the destruction process

If a country has decided to start destruction of a declared part of its chemical weapons, how can this be verified? In other words, how can confidence be built up to confirm that chemical warfare is in the process of being precluded as an option? A number of very relevant working documents on this problem have been produced for the CCD by the United States (USA, 1974, 1976), by the Soviet Union (USSR, 1977) and by Sweden (Sweden, 1976).

At first sight it seems that the actual process of destruction is one of the easiest to get agreement upon for on-site inspections. After destruction, the weapon in question ceases to exist. It has been pointed out by some countries, however, that the identity of the chemical agents to be destroyed might become known via on-site inspection, and that disclosure of this knowledge might lead to the undesirable proliferation of the CW agents and might also infringe industrial rights. This is certainly a point to consider. First, it indicates that very strict security rules and precautions for safeguarding military and industrial secrets must be agreed upon by members of an inspection team. A number of such verification agencies already exist within certain countries or international organizations which verify certain procedures and equipment in the heavily competing chemical process industry.

Second, it is not absolutely necessary to know the exact chemical nature of the compound to be destroyed. Of course, the possibility remains that a country may have a secret CW agent in its stockpile and may wish to destroy it without disclosing its identity. This, however, is a rather remote possibility. By far the largest part of the stockpile will consist of some type of nerve agent (organophosphorus compound or carbamate) or a vesicant (mustard, lewisite). (Dual-purpose agents are not mentioned here as there seems to be no sense in their destruction.) In these cases, owing to different industrial processes it is very likely that countries have chosen different types of these classes of agents. The same applies to the addition of stabilizers, dispersion additives, and so on. If, however, the *type* of agent to be destroyed is declared, it will be possible to verify this by suitable (bio-) chemical means without necessarily having to elucidate the chemical structure completely. In the case of the first-mentioned secret chemical agent, the procedure proposed by Sweden could be applied: verifying toxicity only by injection or by percutaneous methods. As previously stated, this would be an exceptional case.

The best way to verify the destruction of CW-agent stockpiles would be to use multinational regional destruction sites. An example is the German-built Dutch-owned ship *Vulcanus* for destroying industrial chemical wastes. Not only would this method enhance the possibility of confidence-inspiring verification, but it would probably also reduce destruction costs considerably and facilitate the difficult task of disposing of decomposition products. Even when allowance has been made for the fact that some of the objections raised by the Soviet Union in its paper CCD/539 (USSR, 1977) are also applicable to such regional sites, the above-mentioned advantages still constitute ample reason for carefully evaluating the possibilities.

The second-best alternative would be to invite inspectors to national destruction sites. In this case an estimate of the amount of (potential) agent destroyed could be made by the measurement of batch sizes (USSR, 1977) or of flow rates (USA, 1976). If the destruction were carried out by means of a chemical reaction, then the amount of reagent could also be estimated. The aforementioned limitations that environmental control imposes on the entire procedure and that have a tendency to slow down the destruction phase considerably, may, from the verification standpoint, be a blessing in disguise. The virtually total containment of the process makes it possible to establish a material balance, greatly assisting verification.

If the type of agent could be declared, then general (bio-) chemical analytical techniques would give a rather good estimate of the actual amount of agent being destroyed. If this cannot be done, then the method described by Sweden (Sweden, 1976) would have to be adopted.

V. Conclusions

The overall issue of the verification of the destruction of a CW agent stockpile can be subdivided into the following components:

1. The size of the stockpile.

2. The percentage of the stockpile to be destroyed and the rate of destruction.

3. The possibility of confirming the rate of destruction.

An answer to problem 1 can only be obtained by gathering intelligence—satellite observations, estimates of the size of the chemical industry of a state, and so on. As previously stated, in peace-time the stockpile will probably be distributed over a smaller number of well-protected sites, in the same way as tactical nuclear weapons. If this assumption is correct, a reasonable guess of at least the order of magnitude of the chemical stockpile of a nation can be obtained.

The second problem is essentially one of declaration, the rate of destruction depending, of course, on the destruction facilities available to a country. The more closely the quantities of the stockpile destroyed approach the total estimated size of the stockpile, the higher will be the confidence that the country in question is indeed eliminating its chemical warfare potential.

The third issue is the crux of the matter. Destruction carried out at multinational regional destruction sites is the most easily verifiable, but on-site inspection at mutual destruction sites may also create a great deal of confidence with the methods available today. Great care will, however, have to be taken to safeguard the military and industrial proprietary rights of nations. This problem is certainly not insurmountable, as shown by several examples of already existing safeguarding procedures.

References

SIPRI, 1973. *The Problem of Chemical and Biological Warfare*. Volume 2: *CB Weapons Today*. Stockholm: Almqvist & Wiksell.

Sweden, 1976. Disarmament Conference document CCD/485, 5 April.

USA, 1974. Disarmament Conference document CCD/436, 16 July.

USA, 1976. Disarmament Conference document CCD/497, 29 June.

US Dept of the Army, 1977. *Operation of the Chemical Agent Munitions Disposal System (CAMDS) at Tooele Army Depot, Utah*. Final Environment Impact Statement, Office of the Project Manager for Chemical Demilitarization and Installation Restoration, March.

USSR, 1977. Disarmament Conference document CCD/539, 3 August.

Verification problems—monitoring of conversion and destruction of chemical-warfare agent plant[1]

R. E. ROBERTS
Midwest Research Institute, Kansas City, Missouri, USA

Abstract. The interplay between mutual trust, verification and confidence building is briefly considered. The verification implications of converting existing chemical-warfare (CW) agent plants for use in the production of civilian products are described. Dismantling of existing CW plants is suggested to be a less difficult verification problem. Requirements of verifying that declared plants were in fact CW plants are noted. Verification requirements of dismantled and "moth-balled" plants are examined.

I. Introduction

This paper focuses specifically on the verification of the conversion or deactivation of existing chemical-warfare (CW) agent production plants. It assumes that existing CW plants will be declared under the terms of a treaty which bans such production. In this situation the accompanying verification system may have to deal with several degrees of conversion/deactivation. Several possibilities will be examined *vis-à-vis* their verification requirements and the techniques which may be employed to assure compliance.

At the outset the concept of verification *per se* needs to be examined. The nations of the Eastern bloc have often advanced the view that verification systems are not needed to police arms control agreements. Instead, they have argued that mutual trust among nations should be substituted for verification provisions. In truth the reverse situation exists. Verification provisions are required because of the absence of mutual trust. If mutual trust truly existed, there would be no need for arms control agreements or, for that matter, for arms.

Regrettably, mutual trust does not exist in today's world. Given this situation, arms control agreements are attractive means for reducing military expenditures as well as threats posed to one nation by other nations' military capabilities. Verification of compliance with the agreements is the *sine qua non* for expenditure and threat reduction. Interestingly, however, verification systems which are now essential because of lack of trust, can, by the assurance of compliance they can provide, become one of the most powerful tools for building the sought-for mutual trust. Both effects—assurance of compliance

[1] This paper is based on the findings of Midwest Research Institute's continuing programme of research concerned with chemical-warfare verification issues. The research team includes T. Ferguson, A. Meiners, C. Mumma, R. Roberts, D. Rose, and J. Spigarelli.

and confidence building—need to be kept in mind when evaluating verification options.

In sum we have the following equation: the more absolute the verification—born in mistrust—the greater the progress towards absolute trust.

With this preamble to establish context, it is fruitful to turn to an examination of verification issues associated with the conversion or deactivation of CW-agent production plants. Verification confronts several conversion/deactivation options. Inherent in each option are questions which must be answered by a verification system if compliance is to be assured. There is a basic question which underlies each of those geared to a specific situation. It is: How rapidly can a converted or deactivated CW production facility be rededicated to agent production?

II. CW plant conversion

Dual-purpose agents

Strictly speaking, conversion is not an issue. The chemical being produced is a routine item of civilian commerce and has utility as a CW agent, e.g., chlorine and phosgene. The only question which can distinguish between military and civilian production is purpose. An absolute answer can only be provided by monitoring production, transportation and consumption. Detailed reporting of activities and on-site access at the production site, by shippers and by receivers, would be required.

Assessment of the convertibility of single-purpose CW-agent plants to peaceful purposes

Commercial products can be produced in supertoxic, organophosphorus CW-agent plants. Pesticides, plasticizers and fire retardants are the most likely candidates for such production. The primary distinction between an agent plant and a civilian production plant is that the former is overdesigned for production of the latter. Thus, the manufacture of civilian products could be readily accomplished without destroying the features of an agent plant which would be necessary to rededicate the plant rapidly to agent production. This is illustrated by the list of agent-plant criteria set forth at the Fifth Pugwash CW Workshop (held on 17–19 August 1977 at Cologne) which in the opinion of Bayer representatives would characterize an agent plant. They are:

(*a*) isolated location;
(*b*) sealed reaction units;

(*c*) double-walled piping;

(*d*) no roof exhaust fans as in a conventional plant, rather filtered exhaust air systems;

(*e*) remote/automatic process controls;

(*f*) sampling by remote methods from sealed chambers as opposed to open-air grab-sampling from a reactor; and

(*g*) separate waste treatment system or pre-treatment of wastes before introduction into a common waste stream.

Midwest Research Institute's research into the distinctions between civilian plants and agent plants indicates that all of the Bayer criteria are desirable for an agent plant, but that only the safety remote-control features are essential differences. In either event, it is clear that agent plants differ from civilian plants only in the fact that the former would have additional capabilities. They could then be readily employed to manufacture civilian products and just as readily re-employed to produce agent.

The most likely candidate category of chemicals which might be produced in a nerve-agent plant of US design is organophosphorus plasticizers. Domestic manufacturing plants for organophosphorus plasticizers and for organophosphorus nerve agents usually involve closely related chemical processing operations and common requirements for certain raw materials. The process equipment and operations used in plasticizer plants are also geared to handle phosphorus compounds in comparatively large quantities. Also, the properties of many of these plasticizer compounds are essentially the same as those possessed by agent-related substances. These relationships suggest the feasibility of converting some nerve-agent facilities to organophosphorus plasticizer plants.

An agent production facility could, for example, be converted to the manufacture of plasticizer as described below.

Description of plasticizer process

A representative plasticizer, tricresyl phosphate (TCP), can be produced by reacting cresol with phosphorus oxychloride. Hydrogen chloride is formed as a by-product. The basic chemical reaction is:

$$3CH_3C_6H_4OH + POCl_3 \longrightarrow (CH_3C_6H_4O)_3PO + HCl$$

Cresol and phosphorus oxychloride are blended together in a mixer, and the mixture is passed through a heat-exchanger to a reactor. The reaction is carried out in a continuous motion by the use of three reactors in series. The operating temperature maintained in these reactors ranges from 150 to 300°C, and the reactants flow slowly by gravity through each reactor. The hydrogen chloride evolved is passed through reflux condensers to a recovery system.

The completely reacted materials from the last reactor are cooled and treated in a washer with caustic soda solution and water to remove impurities.

A refined grade of product is produced by vacuum distillation; for some uses, the washed material is packaged for sale without further treatment.

Conversion of organophosphorus CW-agent plants to plasticizer plants

The storage tanks for the agent process would serve adequately, without modification, to accommodate the raw materials and product in plasticizer operations.

The reactors employed in the agent process could be altered satisfactorily to conduct the plasticizer process reaction. In order to provide the higher operating temperatures required in plasticizer synthesis, some modifications would be required in the heating and temperature control system. The existing scrubbing equipment could be utilized with only minor changes to permit plasticizer operations. If required, the agent process distillation equipment could be modified for purification of the plasticizer process. The existing heat-exchange equipment could probably be utilized with only minor modifications for the plasticizer process.

Conversion of organophosphorus CW-agent plants to pesticide plants

It would be a relatively easy matter to convert an agent plant to the production of certain pesticides, such as DDVP (dimethyldichlorovinyl phosphate), which is based on the use of phosphorus trichloride, methyl alcohol and chloral as raw materials, and to the production of chlorinated hydrocarbon pesticides.

Conversion of organophosphorus CW-agent plants to manufacture of other civilian chemicals

Although plasticizers and pesticides appear to be the most likely category candidates for agent-plant conversion, there are other product types which could probably be produced satisfactorily in converted agent plants with only minor modifications. Other possible products include phosphorus trichloride and phosphorus oxychloride.

Verification requirements

In those instances where agent plants are actually converted to civilian production, it would generally be an easy matter to reconvert the facility to agent production. Accordingly, the necessary verification procedures are quite stringent. Frequent on-site inspection would be required. The interval between site visits could be established by an initial inspection visit during the conversion process. At that time the inspection team would monitor the physical modifications and review blueprints and other documentation associated with the conversion. It would be important to determine if conversion was actually taking place or whether the modifications consisted only in the installation of a parallel production stream. In any event, the inspectors would determine the time required to rededicate the facility to agent production. That determination would define the frequency of site visits necessary to police the continued dedication of the plant to civilian production. The required interval would probably be measured in days or at best a week or so.

All in all, verification of the conversion of single-purpose agent plants to civilian production seems to demand a level of intrusion which few nations or private companies would tolerate. Additionally, there is little economic incentive to operate agent plants for the production of civilian products. Operating costs would be higher than those of a plant designed from the outset to produce the civilian product in question.

Thus, for verification, economic, and confidence-building reasons it seems much wiser to close down existing agent production facilities completely.

III. CW plant deactivation

Most discussions held at the Conference of the Committee on Disarmament (CCD), at the negotiating body which succeeded it—the Committee on Disarmament (CD)—and at informal forums have centred on a two-step deactivation process: first, a declaration of agent-plant sites; second, the deactivation itself. Thus, there are two sets of verification issues: the first aimed at confirming that the declared plants were in fact agent plants, and the second keyed to verification of agent-plant inactivity under differing possible degrees of deactivation. The rest of this paper will examine four questions integral to these verification issues:

(*a*) a particular facility was designed for CW-agent production;

(*b*) a particular facility had been used for CW-agent production;

(*c*) a given facility has been dismantled and cannot be readily reassembled from its components; and

(*d*) a given facility has been moth-balled and cannot be put back swiftly and clandestinely into agent production.

Guide-lines employed in developing the procedures were that they should be as simple and non-intrusive as is consistent with confirmation that the above conditions exist. Although somewhat different situations will be encountered at plants designed to produce different CW agents or at plants employing different process set-ups to produce a single CW agent, the procedures set forth the types of tests to be employed as well as the types and locations of seals and monitoring devices.

Figure 1 is a display of the general elements of the verification procedures required to confirm each of the four deactivation situations examined in this paper. Narrative descriptions of each of the four situations and the confirmation procedures follow.

Figure 1. Confirmation situations and procedures

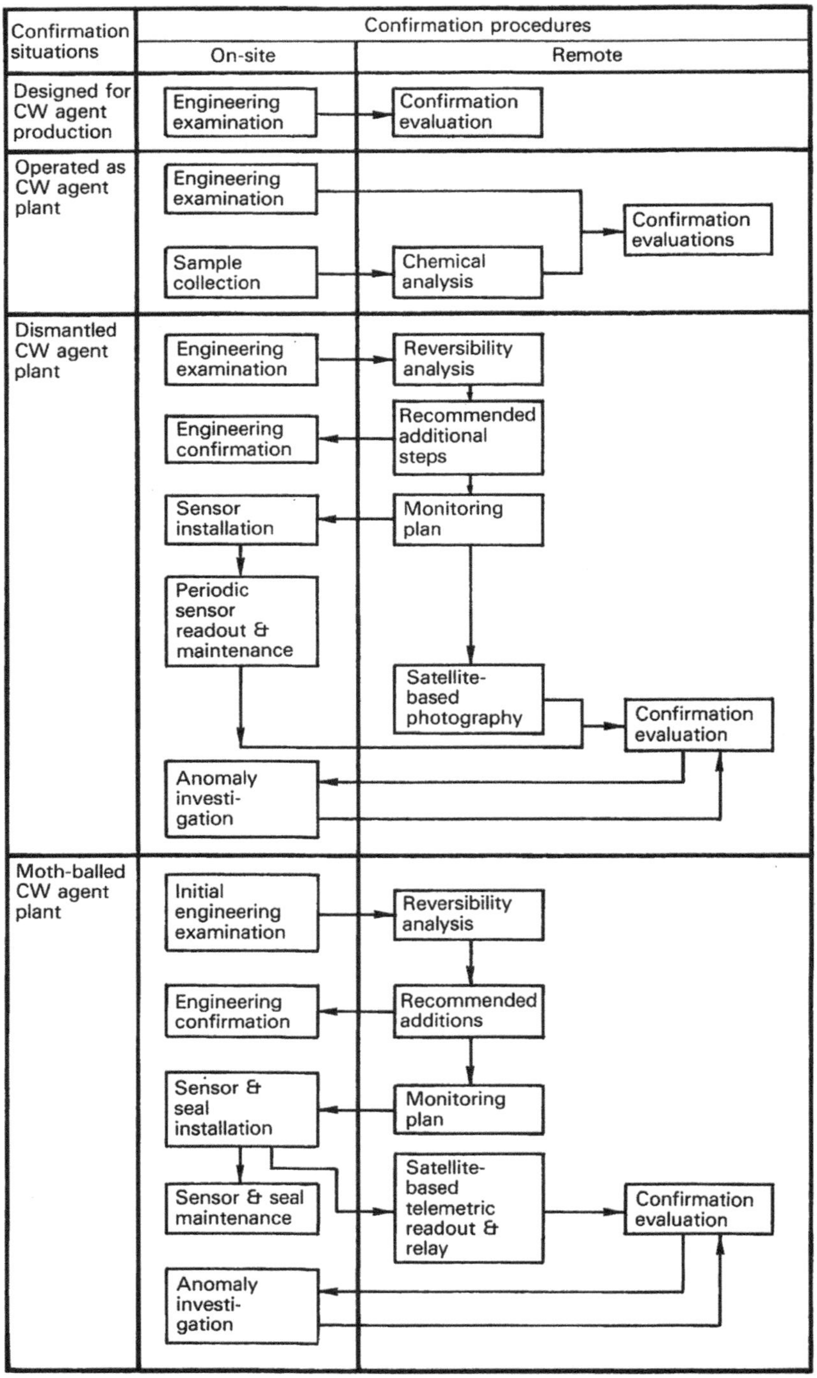

Determination that a declared facility was designed for CW-agent production

Assuming that the plant is inactive, confirmation of this condition is relatively straightforward. It requires a determination that:

(*a*) the necessary chemical processing units, e.g., alkylation, esterification, etc., in the case of organophosphorus agents, are present;

(*b*) appropriate safety features are incorporated into the facility, e.g., remote controls, special storage and filling equipment, and other special safety features; and

(*c*) special waste-treatment equipment is present.

These determinations cannot be made by satellite surveillance or even as near as the plant fence line or building exteriors. In-plant access by a specially prepared chemical processing engineer is necessary.

In the absence of any pre-visit information concerning the agent that the plant was designed to produce, the unit process steps or the input materials, an inspector with appropriate skills can determine that the plant was designed to produce a highly toxic material. Without supplementary information the inspector cannot specify the particular agent or toxic material.

If the on-site inspection is accompanied by documentation detailing the product, flow diagrams, unit processes and their sequencing, input materials and specific waste-treatment processes, a highly reliable determination of the particular agent for which the plant was designed can be made.

In either event no elaborate verification equipment will be required. The major variables are the length of time the inspector must be on site to make the determination—perhaps a week in the case of no supplementary information to less than half as much on-site time with appropriate descriptive documentation—and the precision of the inspector's determination, i.e., whether designed for a highly toxic material or for a particular agent.

Determination that a particular facility had been used for agent production

Confirmation of this condition requires that the questions posed in the preceding case be answered via the application of the postulated procedures; in addition, evidence of agent or degradation products of agent must be obtained. The additional evidence can only be obtained by collection of samples at the site and by subsequent chemical analysis to determine their composition.

Experience in the collection and analysis of samples at US CW-agent and pesticide plants has demonstrated that the product manufactured can be identified from samples collected from within the production equipment, from waste deposits, and from soil in the near vicinity of the plant. However, representatives of Bayer indicated at the Fifth Pugwash CW Workshop that they believed that the technology exists to construct a 'zero discharge' plant. This should pose no problem in confirming, based on samples collected outside the plant, that plants which operated in the past had produced agent, but may become an issue should an agreement not be reached for some years.

To preclude a spurious finding that a plant had been employed in agent production, samples should be obtained from a number of points within the perimeter of the plant. Otherwise, "seeding" might be employed to give a false finding. To offset this ploy, 20 to 30 samples drawn from the equipment and area should suffice.

No elaborate equipment will be required to obtain the necessary samples or preserve their integrity during transportation and storage prior to analysis. The analysis itself will require sophisticated equipment.

Determination that a CW production facility has been dismantled and cannot readily be reassembled

Verification in this instance involves a two-step process. The first is aimed at confirming that the degree of dismantling is such that reassembly cannot occur in a short time period. The second is a continuing monitoring process to ensure that reassembly does not occur. Generally speaking, the greater the initial disassembly, the less frequent or intrusive the continuing monitoring need be.

Major classes of equipment which may be included in the dismantling are: process reaction units (e.g., reactors, distillation columns, inert gas systems and pyrolysis furnaces), process control instrumentation and equipment, safety equipment, waste-treatment equipment and facilities, filling lines, as well as the associated ducting, plumbing, and storage vessels. It is impossible to assess reversibility in the abstract because it can range from complete cannibalization of a site to the removal or plugging of piping links between processing units. Therefore, the first step in the verification process must include an assessment of the reversibility of the type of dismantling which has taken place at a particular plant. This assessment requires on-site access by trained process chemical engineers and chemical plant construction engineers. It cannot be made at the perimeter of the plant or by remote observation.

The assessment of the reversibility of the dismantling is required before a monitoring procedure can be specified which will ensure that the plant is not reconverted to agent production. In the event of major disassembly it may be possible that the second-phase monitoring can be confirmed by satellite observation for indicators of reconstruction combined with the privilege of conducting challenge on-site visits when satellite observation produces indicators of abnormal activity at the site.

In the event of less extensive dismantling which might permit reactivation in a period of weeks, satellite observation should be augmented by the installation of a limited number of tamper-indicating seals on critical items such as control panels and feed lines to reactor vessels. Temperature-sensitive devices can be installed at points in the process which are inherently temperature-dependent when the plant is in production. Seismic devices and pre-programmed cameras can also be employed to detect activity levels greater than those characteristic of a deactivated plant. The integrity of all such sensors can be preserved by the installation of tamper-indicating seals such as the fibre optic seal described in Disarmament Conference Document CCD/498 (USA, 1976).

If on-site sensors are employed, readings can be relayed via telemetry to existing communications satellites or obtained during periodic site visits to maintain the monitoring equipment. The maintenance period varies according to the type of sensor, but generally ranges between four and six months. Thus, when on-site sensors are employed, there will be a continuing requirement for on-site access.

The degree of dismantling and the subsequent time for reassembly are the most important determinants of both the type of monitoring necessary (i.e., satellite observation versus installed on-site sensors) and the frequency of monitoring. Thus, the intrusiveness of the continuing monitoring can be significantly reduced by permitting the inspectors to suggest further dismantling steps in addition to those found during the initial on-site visit. This, of course, would not be necessary if the steps already taken would require something of the order of six months to reverse. However, the option of such suggestions could be useful in developing a simple verification system. This approach would require provision for two initial site visits, the first to check out the degree of dismantling and recommend further steps, if desirable, and a second to confirm that the additional steps had been taken. After that, satellite-based monitoring would suffice to ensure that the plant was not clandestinely reassembled.

Confirming that a declared CW-agent plant had been moth-balled and cannot be clandestinely put back into agent production

The confirmation procedure applicable in this instance is quite similar to that for a dismantled plant. The major difference is that the plant is being preserved intact; therefore the time required to reactivate is inherently shorter.

As before, the confirmation procedure will comprise two phases—the first to confirm that moth-balling has occurred and to install second-phase monitoring devices; the second to continue monitoring in order to ensure that the plant is not reactivated.

Two initial on-site visits will be necessary to phase 1: the first to assess the state of moth-balling and to determine specifically the types of monitoring devices and their location; the second to perform the actual installation of monitoring sensors. The combined time of the two visits could well fall in the range of two weeks to one month, since there will be a significant amount of engineering analysis and a rather detailed engineering and construction effort associated with the installation of the monitoring instrumentation.

Satellite-based monitoring is moderately attractive as a procedure for use in the second-phase monitoring. There are, however, two characteristics of a moth-balled plant which argue against complete remote monitoring. They are:

1. Maintaining a plant in a stand-by condition requires that certain parts of the plant be operated from time to time so as to prevent deterioration. Pumps, for example, must be turned on periodically to ensure that seals retain their integrity and to distribute lubricants throughout the piping. If this is not done, rust and corrosion will defeat the purpose of moth-balling.

2. A plant in a stand-by state also implies a continuing maintenance programme. Thus, there will be some degree of activity at the site. Satellite-based surveillance is ill equipped to detect partial operation of the plant, much less to distinguish between such maintenance operation and a clandestine run. The same is true of the ability of satellite-based monitoring to discriminate between maintenance activity and operating activity. Therefore, acceptable confirmation of non-operation will require the installation of on-site plugs, seals and monitoring devices.

The simplest and most effective set of on-site monitoring devices would be a combination of temperature sensors installed in each of the major reaction units (probably three to five) plus a physical barrier that would deny access to the central control panel. Also motion-triggered cameras and seismic sensors could be used effectively to detect activity in critical areas. The integrity of both the sensors and the barrier would be preserved by the installation of tamper-indicating seals.

Access to the plant would be required on a continuing periodic basis to service the installed devices. Telemetric relay of monitoring readout is a more attractive option in this case than in the preceding case. During the early years of a moth-balled state, a plant could probably be put into operation in several days to a few weeks. (Bayer indicated that it takes them three days to start up their pesticide operation following the annual August vacation shut-down.) When telemetry is employed, access would be limited to that required for instrument maintenance per year. Provision for challenge visits must be incorporated into the procedure. Anomalous readings from telemetered signals would not necessarily indicate a violation; on-site visits would be necessary to confirm the actual cause.

Reference

USA, 1976. Disarmament Conference Document CCD/498, 29 June.

Confidence-building measures and a chemical weapons convention

S. J. LUNDIN[1]
National Defence Research Institute, Department 1, Stockholm, Sweden

Abstract. The international discussion on confidence-building measures (CBMs) is reviewed. Possible CBMs in connection with a future chemical weapons ban are discussed. Their application before and after the entry into force of a convention is treated specifically. Some suggestions on possible CBMs are presented and their relations to possible future verification measures are discussed.

I. Introduction

The explicit concept of confidence-building measures (CBMs) originated in the deliberations of the Conference on Security and Co-operation in Europe (CSCE). In this context, CBMs have been established and regulated in the Helsinki Declaration. Their application and possible subsequent extension are fairly widely discussed in the literature (e.g., Holst and Melander, 1977; Pugwash, 1978; Alford, 1979 *a*, 1979 *b*; Frye, 1979; Lundin, 1979; SIPRI, 1979). Further discussions may take place in the preparatory states for, and during, the second follow-up CSCE session to be held in Madrid in 1980. Political statements also point in this direction (e.g., Brezhnev, 1979).

The concept of CBMs has also been discussed to some extent in the Mutual Force Reduction (MFR) negotiations in Vienna. CBMs are there referred to as "associated measures", and no particular trends or views seem to have emerged so far (Alford, 1979 *b*).

In the past few years the concept of CBMs has also appeared in another arms controls context, namely, in the work of the Committee on Disarmament (CD) in Geneva—formerly the Conference of the Committee on Disarmament (CCD)—for reaching a prohibition on the development, production and stockpiling of chemical weapons and to prescribe their destruction (Sweden, 1979).

The trust that must be generated between nations before they can reach international binding arms-control and disarmament agreements is naturally also required for reaching a ban on the acquisition of chemical weapons. The first proposal for specific CBMs to promote trust as regards abolishing chemical weapons was the draft convention on chemical weapons presented

[1] Since the author is a Swedish civil servant, it is necessary to state that the opinions expressed in this paper are his alone and do not necessarily reflect the views of the Swedish government or of other Swedish authorities.

by the British representative to the CCD in 1976. Since then the notion of CBMs in connection with a chemical weapons convention has won increased attention both in the CD (Sweden, 1979) and elsewhere (Pugwash, 1978; Lundin, 1979; SIPRI, 1979).

This paper discusses CBMs *vis-à-vis* chemical weapons, emphasizing their possible usefulness in connection with the disposal of chemical weapons or the conversion of stockpiles of chemical weapons and their production facilities to peaceful purposes.

II. General

Before entering into a general characterization of CBMs, two differences between possible CBMs in the chemical weapons field and those discussed in the CSCE context should be pointed out: (*a*) the latter always relate to military matters, while (*b*) the former may also include measures relating to civilian matters. To the extent that CBMs are unrelated to military security, they may be the more easily instituted and accepted. On the other hand, they may give rise to other problems of a technological, industrial or economic nature.

A second general feature relates to the different nature of CBMs before and after the entry into force of a convention. The main difference is that pre-convention CBMs are necessarily always voluntary, but those instituted under a convention—apart from possibly being voluntary, as under the Helsinki Declaration—may also be obligatory, which may be an option for a future chemical weapons convention. Post-convention CBMs may also possess another characteristic—it may not be possible for various reasons to verify the information provided, and this would call for unconditional belief in its veracity. This problem will be taken up again in this paper.

Some thoughts have been expressed about the intentions behind CBMs. Thus, with respect to the military situation in Europe, Holst and Melander (1977) note that "confidence building involves the communication of credible evidence of the absence of feared threats". They state, however, that CBMs only add incrementally to information obtainable by intelligence of various kinds. They also emphasize that CBMs do not aim at a direct reduction in rival military efforts, but that they ought to provide reassurance to states by reducing uncertainties and by constraining opportunities for exerting pressure through military activity. This last-mentioned aim is not in line with the working of a chemical weapons convention, the aim of which is in fact to abolish a military potential—that is, chemical weapons.

Alford (1979 *b*) remarks that, in contrast to arms-control verification methods—which according to him concern qualitative and physical characteristics—CBMs operate on the perceptions and intentions of those in confrontation. He further states that CBMs could be continuous in order to demonstrate non-aggressive postures publicly. An example would be future "associated measures" in MFR in monitoring agreed ceilings and deployments by military forces. Another type of CBM would be in operation in times of crisis, making parties less vulnerable to a surprise attack since they would be assured of warning. Alford points out that CBMs are very slow-acting and that confidence will not be acquired overnight.

Alford (1979 *b*) also remarks that it would be unproductive to classify weapons as either offensive or defensive. This may be true provided that one is speaking of the *use* of weapons. However, in discussing a weapons system—such as chemical weapons—for total prohibition, the terms offensive and defensive have to be used with other connotations. Any use of the weapons must then be characterized as offensive, in the sense that they are launched in order to inflict damage on the adversary. On the other hand, all measures aimed at reducing or eliminating the effects of attacks by chemical weapons may be characterized as defensive. In the case of chemical weapons, few defensive measures can be mistakenly taken to be offensive. However, offensive and/or defensive—or rather 'protective'—measures constitute what may be termed a chemical warfare (CW) capability (Pugwash, 1978; Lundin, 1979). It should be borne in mind that protective measures are intended to be allowed under a future chemical weapons convention.

Further, the use of chemical weapons is already explicitly prohibited by the Geneva Protocol of 1925, so that the problem of offensive or defensive use should, theoretically, not exist. However, owing to reservations to the Protocol, in effect it only prohibits the first use of chemical weapons. The defensive measures, although they have to be coordinated with the protection against conventional weapons, are to a large extent specifically aimed at the physical and physiological actions of the agents themselves. Moreover, effective protection against these weapons is comparatively easy to attain. The incentive to use chemical weapons against an adequately protected adversary might be relatively small. The development and application of protective measures may imply readiness for resisting an attack by chemical weapons as well as serve as a prerequisite for a capacity to wage offensive chemical warfare. Thus, as long as a party can maintain and demonstrate effective anti-CW protection and show at the same time that it is not making, or preparing for, offensive use of chemical weapons, it both stabilizes the situation by diminishing the incentives for the use of chemical weapons against itself and increases the confidence on the part of the other parties that it does not itself intend to use them.

The concept of stabilizing and destabilizing uncertainties has been treated by Frye (1979) in the SALT (Strategic Arms Limitation Talks) context. He notes that the stabilizing uncertainty created during peace-time by the question of whether or not nuclear weapons may be used in war may change into a destabilizing uncertainty in a time of crisis, when the need for a pre-emptive strike may be strong for the parties involved. This situation may not be strictly true in the case of chemical weapons. The most destabilizing factor, in fact, is the uncertainty as to which countries possess chemical weapons. The USA is the only country that at present admits to possessing them. Some countries have declared their non-possession of chemical weapons, and also that they do not intend acquiring them. In other words, neither capabilities nor intentions are as clearly expressed as they seem to be in the SALT context.

This situation of uncertainty clearly derives from how parties assess different means for best preserving their security. Different conditions in different countries naturally result in differing judgements on the appropriate degree of openness. However, a particular problem arises when a number of parties—each practising a degree of openness suited to its normal military capabilities and overall security aspects—have committed themselves to reaching a prohibition of a particular weapons system, such as chemical weapons. In this case, information on capabilities and intentions constitute stabilizing CBMs, with little or no revelation of facts not pertinent to the agreed provisions. Prolonged absence of information may bring about destabilizing effects by diminishing the confidence in a party with respect to its intentions. A particular problem in this connection may arise from the failure or unwillingness of parties to take note of or to participate in CBMs investigated by other parties. In the context of voluntary pre-convention CBMs it is not a party's right to abstain from noting or participating in such CBMs that *per se* gives rise to the problem. It is rather the unfavourable effect of this negligence on the CBM-offering party that may cause a backlash—a destabilizing sense of disappointment. Thus, the manner in which CBMs are noted, accepted or reciprocated greatly influences their effect and value. These conditions point to the need for careful analysis of stabilizing or destabilizing effects of individual CBMs.

The question of implementation of voluntary and obligatory CBMs is partly linked with their application before or after a convention is in force. Pre-convention CBMs must be regarded as voluntary, irrespective of the conditions that might influence a party to implement them. The reason for implementing them may be to alleviate perceived threats by other countries, irrespective of whether a convention is being negotiated or not. If such negotiations are under way, an additional reason might be to facilitate the negotiations.

The situation is different under a convention. Voluntary CBMs may also occur under a convention, whether it provides for them or not. But the concept of CBMs may be broadened somewhat when instituted by a convention in that such CBMs may not only be voluntary but also obligatory. In this case, it might be preferable simply to call them convention provisions rather than confidence-building measures. The main reason for not doing this is that some obligations under the convention—although necessary to make a convention trustworthy—may not, for various reasons, be subject to verification. Under these circumstances, information provided by a party's national authorities may have to be accepted at face value without (effective) verification and would, accordingly, be better referred to as obligatory CBMs.

The need for obligatory CBMs may accordingly be particularly obvious for a chemical weapons convention with its extremely complicated relations between civilian and military conditions. Normally, trust and confidence in the observance of a convention may prevail as long as obligatorily requested information is continually provided by parties. As a matter of fact, a continuously expressed commitment to a cause—in the form of a continual supply of information on the matter—may make it politically difficult for a country to violate a convention even if intrusive verifications of the particular issue might not be technically possible to undertake. Further, nationally provided information, perhaps provided over a long time, should also be useful if intrusive international control could be instituted by means of complaints to a consultative committee (Lundin, 1977; Alford, 1979 *a*) under the convention or to an international disarmament organization (Myrdal, 1974).

It must be emphasized that the possibility of instituting and accepting obligatory CBMs should not substitute for international verification measures. They should be contemplated particularly only when agreed intrusive verification may not be technically feasible under a convention. With respect to a chemical weapons convention, the concept will be discussed below with regard to stockpiles of chemical weapons.

III. CBMs for a chemical weapons convention

General

The foregoing discussion indicates that CBMs related to a CW convention involve more applications than have so far been discussed, particularly in connection with the CSCE, where they relate solely to military matters. However, it is not the intention to dwell here in any depth on the classification aspects. The concepts discussed will be applied in a rather pragmatic way. Since the discussion concerns a convention to be, it will relate to the conditions before and after the coming into force of a possible convention. The latter part of this paper does not imply any preferences on the part of the author as to what such a convention should prescribe, but is only intended to be illustrative. Further, opinions might differ as to whether some of the given examples really constitute CBMs, but it is beyond the scope of this paper to enter into a general discussion on such matters.

Pre-convention CBMs

It may be advantageous to bring about a general discussion among negotiators of a chemical weapons convention of the possible value and practical applications of pre-convention CBMs. The above-mentioned risk of backlashes may thus be diminished. In this connection it should be borne in mind that instituting CBMs is not one of the main tasks of negotiators, and that the application of CBMs can in no way substitute for negotiations resulting in a convention. However, in so far as CBMs may facilitate and perhaps shorten the time for negotiations, they deserve a thorough analysis. It is accordingly suggested that the CD in Geneva devote some time to explicit discussions on CBMs in the chemical weapons context. Apart from leading to a possible recognition of the value of CBMs, such a discussion could also lead to the identification of some specific CBMs that some negotiating parties might find worthwhile to pursue. It should be noted that some countries have already undertaken such CBMs both in the CD and in other connections. By way of example may be cited the two visits recently arranged to chemical production facilities in the Federal Republic of Germany and the United Kingdom (UN, 1978; UK, 1979 *a*, 1979 *b*). Similar visits as well as a visit to a chemical weapons destruction facility were carried out by the Pugwash Chemical Warfare Study Group (see footnote 2 on page 1) Workshop in FR Germany and the USA during 1977 and 1978, respectively (Lundin, 1977; SIPRI, 1979). These visits showed that, apart from their confidence-building value, they also provided excellent opportunities for experiencing and discussing the difficult technical problems that have to be solved in connection with a production ban on chemical weapons.

With these previous undertakings in mind, the following suggestions are offered on possibly suitable CBMs to be applied during the negotiation phase for a chemical weapons convention.

Visits to chemical industries

The visits hitherto carried out have aimed at illustrating that technical experts can visit chemical production facilities to see for themselves that nerve agents are not being produced there. This can apparently be done without jeopardizing industrial secrets to the detriment of the companies involved. As mentioned, the visits in question also threw light on other problems of the chemical industry *vis-à-vis* a ban on the production of chemical weapons. A particularly important problem in this respect is the growing increase in safety measures in the chemical industry, since this may diminish the value of the absence of safety measures as an indicator of non-production of CW agents. The problem may also be mentioned of how the verification of the production of former (now dual-purpose) CW agents, such as phosgene, or of toxic insecticides, say, for civilian purposes, could possibly be addressed as a CBM. A third area relates to the subject of conversion of CW agents to peaceful uses. Two types of CBM are conceivable. A country producing CW agents[2] either in special plants or within the facilities of a larger chemical-production area

[2] Unless otherwise stated, this paper deals with single-purpose CW agents (Lundin, 1975).

might declare this fact. Once these conditions are known, one might declare or demonstrate plans or facilities for either destroying or converting the CW agents to peaceful purposes to be undertaken under a chemical weapons convention.[3] The economic and technical aspects of destruction and conversion are discussed elsewhere in this volume (e.g., the papers by Lohs and Mikulak, on pages 67–75 and 57–66 respectively). Features of these problems may vary from country to country, depending on conditions, such as the existence of controlled or free-market economies and the stringency of environmental control regulations. The unilateral provision of information on such special conditions would probably serve as a CBM. Some attempts in this direction have already been made in various connections, for example, in the CD.

Declarations on possession of chemical weapons

The most important type of voluntary pre-convention CBM today would seem to consist in supplying information on possible possession of chemical weapons or on intentions concerning non-possession of such weapons. In the beginning of this paper the stabilizing effects of CBMs were discussed in general—for example, what effect knowledge of the existence of nuclear weapons in a country or in an area might have. Declarations on possession of chemical weapons would not parallel similar declarations on nuclear weapons. First, chemical weapons are not strategic weapons, at least not for the two main military blocs. Second, it is impossible to deal a pre-emptive blow against the chemical weapons of an adversary. Third, the diminished effects of chemical weapons owing to the use of protective measures in military conflict does not find a parallel in the case of nuclear weapons. Thus, the military importance of the knowledge that a country possesses chemical weapons would be rather small. Only in an area where virtually no protection against chemical weapons exists would uncertainty about possession of chemical weapons play any particularly important role. On the other hand, reluctance to declare possession might cause serious suspicion and distrust regarding the general intentions of such a party, at any rate as long as it takes part in negotiations on abolishing chemical weapons. In other words, non-declaration may have a destabilizing effect. As long as it results simply in an increase in protective capacity by other parties in the area, the destabilizing effect may not become dominant. However, as soon as non-declaration also leads to an increase in offensive CW capability, a situation arises that may be difficult to assess—in other words, a destabilizing situation sets in.

Co-operation on protection

In view of the deterrent and thereby the stabilizing effect that well-developed protection against chemical weapons may exert on potential adversaries, one

[3] For more thorough examples of sequential alternatives, see Lundin (1979).

might argue that international co-operation and increased protective measures might be confidence building, as would internationally organized assistance to victims of chemical attack. See, for example, Yugoslavia (1976 *a*, 1976 *b*) and UK (1979 *b*).

However, there may also be arguments against increased protection and co-operation in this field. The one heard the most often is also the most debatable one. It advocates the total abolition of protection. Again, this view derives from the strategic debate on nuclear weapons. It relates to the well-known concept of Mutual Assured Destruction (MAD), this being the ultimate deterrent which will prevent the nuclear powers from even starting a nuclear war. For this concept to work, it is necessary that neither party can or will provide protection for its civilian population. Any attempt at doing so is interpreted as an attempt to survive the opponent's first strike and thus enable oneself to deliver one's own second strike. Similarly, it is often argued that satisfactory protection against chemical weapons is also a sign of an existing or planned offensive CW capability. Thus, good or increased protection might invite other parties to procure chemical weapons of their own, and would thereby be destabilizing. It is here that the parallel with nuclear weapons is misleading. Chemical weapons are mainly tactical and not strategic weapons. Thus, their possible use cannot hold hostage the populations of entire continents. Accordingly, the absence of protection does not increase the deterrent effect. On the other hand, foreseeable tactical applications of chemical weapons require both the attacking and the attacked parties to be protected; the protective measures, in addition to counteracting the effects of chemical weapons, also become very cumbersome for the normal waging of war for both parties, both factors contributing to the deterrent effect of protection. In addition, there is the need to protect the civilian population from the collateral effects of the tactical use of chemical weapons. At least this is important in countries which might suffer attack with chemical weapons on their own soil. However, it is true that an offensive capability must include satisfactory protection to be of any significant military use. This does not detract in any way from the deterrent effect of the protection *per se* of a party, as long as it can convince others that it has no offensive CW capability and neither does it wish to acquire one.

Another argument states that development of protective measures implies an advanced knowledge of the chemical weapons themselves. This might also lead to a precautionary development of increased knowledge of new chemical weapons and also to proliferation of chemical weapons to states with less advanced knowledge in this field. In particular, the dissemination of knowledge about CW agents against which the protective and therapeutic measures are not effective might have a proliferating effect. In view of the fact that existing chemical weapons are already effective enough, and that agents of much lower toxicity may be used for retaliation or even for attacking purposes, the extra proliferation risk of international co-operation on protection should not be exaggerated. One can safely draw the conclusion that such co-operation would in fact be stabilizing and confidence building.

Monitoring scientific and technical development

The foregoing reasoning on protection does not exclude the risk for new developments in the field of chemical weapons. On the contrary, disturbing scenarios exist where chemical effects are used both directly and in combination with other types of weapon. It thus seems necessary to follow scientific and technical developments by monitoring the scientific and technical literature, for example. Several suggestions in this direction have been made in the CCD and CD (e.g., Sweden, 1978). It is difficult to envisage any common international effort for such monitoring before a convention has entered into force. Further, purely scientific activities are not intended to be prohibited under a convention; only those technological developments that are aimed at developing CW agents are envisioned for prohibition. Thus, international monitoring of the literature may never be instituted under a convention. However, countries would most certainly during a foreseeable future need to undertake national monitoring activities in this direction. Accordingly, international exchange of information and discussion of monitoring methods might be confidence building (Sweden, 1978).

CBMs under a chemical weapons convention

General

As explained in the introduction, some obligatory measures or provisions under a convention may also be characterized as CBMs when the pertaining information must not or cannot be verified by international verification measures. Some such CBMs may have to be accepted under a final convention. It is not intended in this paper to speculate on the activities that might have to be internationally verified in a future convention. Only a couple of possible examples of each category will therefore be given.

Voluntary CBMs

In other connections it has been argued that a future chemical weapons convention, in order finally to abolish the threat of chemical weapons (Lundin, 1979; SIPRI, 1979; Sweden, 1979), ought to include prohibitions of the planning, organization and training activities necessary for an offensive CW capability. Initially, at any rate, such prohibitions may be very difficult to verify, since they concern purely military matters. However, in this connection, note should be made of the precedent on CBMs that has been set in the Helsinki Declaration and that may be followed by similar measures if and when the MFR talks lead to an agreement.

In addition, the chemical weapons convention might to some extent apply the same system as the Helsinki Declaration in the above-mentioned context. Thus, parties could invite military observers to manoeuvres where training in anti-CW protection was being practised. Other observers might be able to visit NBC (nuclear, biological, chemical) protection schools. One might even conceive of exchanging certain military manuals.

One interesting question would be the regional aspect. The above-mentioned CW activities (planning, organization and training) may not occur or be observable in the manoeuvres which are notified and to which observers are also invited sometimes under the Helsinki Declaration. Thus, one might identify particular facilities that might be visited—for example, NBC schools, as mentioned above. Another, certainly more unconventional suggestion might be for random visits to munition depots.

Obligatory CBMs

Verifiable measures. It is appropriate under this heading to cite examples of CBMs concerning conversion activities.

One might visualize a convention stipulating that disposal of production facilities and stockpiles need not be subject to intrusive international verification in cases where a party wishes to convert the facilities and weapons to peaceful uses by methods that would require secrecy in order not to jeopardize industrial know-how. According to the line of reasoning pursued in this paper, information given without opportunities for verification must be characterized as CBMs. Thus, in times of lowered trust between parties mere CBMs on disposal of chemical weapons will most likely be insufficient. Satellite monitoring would seem to be of very limited use. One might thus hope that some international verification would also be possible in the case in question. If that is impossible, one might instead find it necessary to demand more extensive information than would otherwise be required (see also p. 143). Examples of such items of information may be the following:

- give reasons why the material has to be converted instead of being destroyed;
- announce amounts of CW agents to be converted and the time-schedule for the activity;
- announce the facilities where the conversion shall take place; and
- announce from where the material will arrive at the conversion facilities, i.e., identify stockpile locations.

Non-verified measures. The convention prohibiting the development, production and stockpiling of biological weapons has no provisions for international verification measures. It does not ask for information on the destruction of possible stockpiles of biological weapons. These conditions seem to be acceptable to the parties to the convention, despite the fact that some states expressed serious concern in this respect during the negotiation of the convention. However, the concern about the need for verification under a chemical weapons convention is more widespread.

One question in particular has been discussed by some of the negotiating parties in the CCD and CD: How is one to be confident that possibly required information on existing stockpiles or on completed destruction of all stockpiles is correct? Such information is technically difficult to verify, at least in very large countries, and perhaps also in countries with extensive industrial areas. In view of this fact, the amount of information given and the conditions under which it is given may be decisive for the confidence that it will generate in other parties. This would probably imply, for example, that reluctance to give any information at all would give rise to distrust rather than confidence. On the other hand, a country which had built up extensive protection against chemical weapons but which did not possess chemical weapons might encounter difficulties in the way of getting information to that effect believed. Other such situations can also be envisaged. Most industrial countries—at any rate those having chemical industries—would have to consider the implications in this respect from their production of dual-purpose chemical agents, that is, chemicals of such toxicity and properties that would make them suitable for use as CW agents. To this category belong chemicals such as phosgene and hydrogen cyanide, which were used as CW agents during World War I.

However, it is obvious that the serious problem here concerns the possibility of hiding—in violation of a convention—chemical weapons ready for use at short notice (and to some extent the production facilities for them), and the fact that no short-term solution to this problem exists today. A thought-provoking discussion on the subject is given by Ooms in his contribution to this book (pages 123–128). However, a ban including the prohibition of planning, organization and training would have long-term effects in this respect. It seems highly unlikely that this uncertainty would jeopardize a future convention as long as steps were taken over the whole field by every party to reassure all other parties about their serious intentions to adhere to the provisions of a convention. Such steps would include particular measures to make information on a party's stockpiles appear acceptable. Some such measures are CBMs, as previously stated in this paper. They have conceivably to be reinforced by different kinds of verification measures. As an intermediate step, a measure may also be utilized under a chemical weapons convention—a complaints procedure of the type already implemented by the SALT I agreements, namely, utilization of a consultative committee (see also Lundin. 1977).

IV. Conclusions

The concept of confidence-building measures has already been introduced and discussed in arms control connections, such as the Helsinki Declaration and the MFR negotiations. CBMs of different kinds seem particularly apt for application in such a complicated convention as a future chemical weapons convention will apparently be. This may be due to a possible need for reconciling opposing views as to where intrusive international verification measures can be undertaken, and on the usefulness of national means of verification as well as on the capacity of national technical means of verification. CBMs may be indispensable in the case of some activities that may not be verified for various reasons. This especially concerns information on the existence of stockpiles of chemical weapons and production facilities for them. The application of CBMs in these matters needs to be analysed further, so as to make them effective and to prevent them from having an unintentional opposite effect.

CBMs might find particular use in connection with a possible future need to convert chemical weapons and agents for peaceful purposes in preference to destroying them.

References

Alford, J., 1979*a*. Introduction. In *The Future of Arms Control: Part III. Confidence-Building Measures*. Adelphi Paper No. 149, Spring 1979, pp. 1–3. London: The International Institute for Strategic Studies.

Alford, J., 1979*b*. Confidence-building measures in Europe: the military aspects. In *The Future of Arms Control: Part III. Confidence-Building Measures*, Adelphi Paper No. 149, Spring 1979, pp. 4–13. London: The International Institute for Strategic Studies.

Brezhnev, L. I., 1979. Statement of 2 March.

Frye, A., 1979. Confidence-building measure in SALT: a PAR perspective. In *The Future of Arms Control: Part III. Confidence-Building Measures*, Adelphi Paper No. 149, Spring 1979, pp. 14–22. London: The International Institute for Strategic Studies.

Holst, J. J. and Melander, K. A., 1977. European security and confidence-building measures. *Survival*, **19**: 146–54.

Lundin, J., 1975. Description of a model for delimitating chemical warfare agents in an international treaty. In *FOA Reports*, Vol. 9, No. 4, June, pp. 1–10. Stockholm: Research Institute of Swedish National Defence.

Lundin, S. J., 1977. Verification of a ban on chemical weapons—a suggestion for mutual on-site observations. In: Report on the Fifth Pugwash Chemical Warfare Workshop, 17–19 August.

Lundin, S. J., 1979. Chemical weapons—too late for disarmament? *The Bulletin of the Atomic Scientists*, **35**(10): 33–37, December.

Myrdal, A., 1974. The international control of disarmament. *Scientific American*, **231**: 21–33.

Pugwash, 1978. Report on the Sixth Pugwash Workshop on Chemical Warfare in the USA, 8–12 May. *Pugwash Newsletter*, **16**: 4–12.

SIPRI, 1979. Stockpiles of chemical weapons and their destruction. In *World Armaments and Disarmament: SIPRI Yearbook 1979*, pp. 470–89. London: Taylor & Francis.
Sweden, 1978. Disarmament Conference document CCD/569, 24 April.
Sweden, 1979. Disarmament Conference document CD/PV.29, 24 April.
UK, 1979 *a*. Disarmament Conference document CD/PV.2, 24 January.
UK, 1979 *b*, Disarmament Conference document CD/15, 24 April.
UN, 1978. Document A/S-10/23, Part II, 30 June, p. 29.
Yugoslavia, 1976 *a*. Disarmament Conference document CCD/503, 5 July.
Yugoslavia, 1976 *b*. Disarmament Conference document CCD/PV.714, 22 July.

Assessment

SIPRI

SIPRI has paid close attention over the years to problems of chemical-warfare (CW) disarmament. Based on this experience, we summarize here the most salient features, as we see them, of the papers presented in this volume and of the discussions at the symposium from which the papers originated.

The chemical weapons convention now being sought in the intergovernmental fora must provide for the elimination of existing stocks of CW weapons and the means at present dedicated to their production. Our symposium was convened for the purpose of exploring specific problems relating to this basic objective. In particular, important questions had been raised in international discussions about how elimination should best be performed, about the period of time which the convention should allow for it, and about the requisite control procedures. On these matters SIPRI's assessment is now as follows:

1. From the past experience of states that have discarded their capabilities for waging chemical warfare or that have eliminated obsolete, surplus or otherwise unwanted chemical weapons, there exists a substantial body of knowledge applicable to the implementation of chemical disarmament. For those CW agents in respect of which there has as yet been no such experience, current knowledge in chemistry is sufficient to indicate ways in which detoxification could be performed effectively. It is reasonable to conclude, therefore, that adequate means and methods are available, or can be developed, for the elimination of all CW agents.

2. The same past experience has demonstrated, sometimes at great cost, that disposal of CW agents and munitions can create grave toxic hazards to human populations and other forms of threat to ecosystems. Measures adopted to reduce these dangers inevitably increase both the expense of disposal and the time which it would require. It has been estimated that the detoxification of a militarily significant stockpile of CW agents, and the demilitarization of the associated munitions, would take a period of at least several years to complete when performed in accordance with stringent standards of industrial hygiene and environmental protection, and within a budget of realistic proportions. We see nothing unreasonable in such estimates.

3. With respect to the financial burden of elimination, we note that the cost of maintaining stocks of CW weapons safely and securely would also be substantial.

4. Facilities in which CW agent has been manufactured or filled into munitions will inevitably have become contaminated with the agent. Their destruction or dismantling will therefore present public-safety and environmental-protection problems comparable in kind, if not in degree, with those arising from agent detoxification. The period of time required for the

elimination of plant for CW agents and munitions will therefore also be extensive.

5. It is important to appreciate that one consequence of the protracted nature of these elimination processes is that, even after entry into force of the chemical weapons convention, some parties will continue to have a capability for chemical warfare at their disposal for a substantial period.

6. In view of national differences as regards current practice and domestic legislation in the fields of industrial hygiene and environmental protection, and in view also of the differences between countries as regards natural environment, population-density patterns and industrial structure, the final choice of method to be adopted for elimination must be left to the individual countries concerned. Other countries will, however, have certain legitimate interests in that choice. Provisions are therefore required for an appropriate degree of international consultation prior to selection of elimination methodology.

7. Above all, both the choice and the application of elimination technique would require international co-operation to meet the need for adequate verification. Some types of elimination procedure may require less international monitoring than others, especially by states possessing sophisticated 'national technical means' of verification. The choice of elimination methodology should be such as to keep the role of on-site inspection to a minimum within the overall disarmament regime.

8. International consultation and co-operation on elimination would also be prudent as regards the public-safety and environmental-protection aspects of elimination technique. Historical experience in disposal of CW agents and munitions relates, for the most part, to a period in which the applicable standards in both areas were considerably lower than they are today. Moreover, some types of CW agent appear to possess toxicological and ecological properties which are not yet fully understood. Consequently, the full magnitude of the risks to human health and the natural environment that might be created by elimination processes is still unclear. These are risks that may not be confinable within national borders. We note that an added benefit of international consultation on these matters, if continued over the period required for elimination, would be the opportunity for building confidence among states in the implementation of chemical disarmament.

9. Conversion of CW agents or their means of production into products or facilities whose applications are unrelated to chemical warfare represents an alternative to destruction as a method of elimination. From an economic and social standpoint, such an alternative could in some instances present very definite attractions, particularly if the conversion could somehow be made to satisfy specific development needs. This is an option which clearly requires closer study than it has yet received, including, in the case of CW agents, laboratory study of the novel chemical processing that would be involved.

10. Pending such further examination, it appears to us that any economic rationale for choosing the conversion rather than the destruction route would

be strong only in cases where the product of conversion closely resembled the subject of conversion. One such example might be the adaptation of chemical plant to produce civilian commodities similar in chemical structure or toxicity to the CW agent for which the plant was originally built. If this supposition of ours proved correct, it would mean, first, that the economically viable conversion possibilities would be relatively few in number, and second, that conversion would necessitate more intrusive methods for verifying elimination than would destruction.

11. In the case of CW agents, the main reason why we suppose that economic factors would generally militate against the conversion alternative is that large-scale application of chemical-process technology to highly toxic starting materials would inevitably demand heavy capital investment in safety precautions and other containment measures. Thus, while it may be possible in theory to convert nerve gases, for example, into pesticides, plasticizers and the like, or to convert mustard gas into synthetic rubber, fewer resources would probably be consumed if such commodities were produced from their conventional starting materials. This conclusion would not necessarily hold for the less toxic CW agents. Nor would it hold for "dual-purpose" agents or for stockpiles of agent starting-materials and intermediates, including binary-munition reactants. In these latter cases, then, the requirements of adequate verification would be the only major constraints on selection of the conversion alternative.

12. The possibility of developing novel conversion processes should not be taken as sufficient grounds for delaying elimination of the means for chemical warfare.

13. The foregoing assessments relate only to elimination of stocks of CW agents and munitions and their means of production. There are several other physical components of CW capability whose elimination under the terms of a chemical weapons convention would also be necessary or desirable. They include chemical-weapons laboratories, proving grounds, storage installations and training facilities. The practical problems of eliminating these components, which, in most cases, would presumably best be done by conversion rather than by destruction, require investigation.

PART III

Status of US–Soviet negotiations for a chemical weapons convention

J. GOLDBLAT
Stockholm International Peace Research Institute

By a letter of 7 August 1979, the USA and the USSR formally transmitted, for the information of the Committee on Disarmament (CD), a joint report on progress in their bilateral negotiations on the prohibition of chemical weapons (for the text of the report, see appendix 1).

These negotiations, the tenth round of which was held in the summer of 1979, are conducted in private, in accordance with the US–Soviet agreement of 3 July 1974, to prepare a "joint initiative" with respect to the conclusion of an international convention dealing with means of chemical warfare.

The 1979 report is more substantive than previous communications on the course of the US–Soviet negotiations, since it identifies, in greater detail, the main areas of agreement and disagreement and specifies questions requiring further study.

On the basis of this report, as well as the explanations given by the representatives of the two powers, and in the light of the discussions held in the CD, the negotiating situation, as of mid-1979, can be described as follows.

I. Scope of the convention

The USA and the USSR are now agreed that a chemical weapons convention should be comprehensive in its coverage. This means that the parties would assume the following obligations:

(*a*) Not to develop, produce, stockpile, otherwise acquire or possess, or retain chemicals for chemical warfare purposes or chemical weapons;

(*b*) Not to transfer to anyone the means of chemical warfare, and not to assist, encourage or induce others to carry out the activities prohibited by the convention;

(*c*) To destroy or divert for permitted purposes the existing stocks of the relevant chemicals and weapons; and

(*d*) To destroy or dismantle means of production of chemical weapons.

The prohibition of the use of chemical weapons is not specifically mentioned. Since such a prohibition is already included in the 1925 Geneva Protocol, the two powers are prepared to refer to it in a chemical weapons convention, but are unwilling to duplicate the provisions of the Protocol.

The scope of the prohibition would be determined on the basis of a general purpose criterion. This means that the prohibition would apply to chemical substances (which are already in existence or which may be discovered in the future) that have no justification for peaceful purposes, and to weapons specifically designed to use such substances for chemical warfare purposes. More precisely, the ban would cover supertoxic lethal chemicals (such as nerve gas), other lethal chemicals (such as phosgene), and highly toxic but not necessarily lethal chemicals (incapacitants), as well as precursors (including components of so-called binary chemical weapons[1]). It would also cover chemical munitions or other means of chemical warfare (such as spray tanks), with the exception of munitions used for smoke dissemination.

Not all the issues relating to the scope of the convention have been solved. Neither have the terms, referred to above, been defined with a precision required in an internationally binding document. It is, nevertheless, clear that comprehensive coverage is not tantamount to absolute prohibition. Dual-purpose chemicals, that is, chemicals useful for chemical warfare but intended for non-hostile purposes, of types and in quantities which can vary considerably from country to country but are appropriate for these purposes, would be exempted from the ban. It is not clear, however, who would be authorized to judge whether the quantities produced or retained by individual states were justified by "non-hostile" needs.

The permitted purposes would include civilian industrial production, peaceful scientific and medical research, and domestic law enforcement, as well as development and testing of means of protection against chemical weapons. They could even include military purposes, as in the case of production of missile and torpedo fuels which, although toxic, are not related to chemical warfare. In addition, the United States considers that certain specific military uses of chemicals, such as the use of tear-gas for riot control in prisoner-of-war camps, or the use of anti-plant agents for clearing vegetation around the party's own military bases, should also be allowed.

The notion of permitted and non-permitted purposes implies that different degrees of prohibition and limitation would be applied. Consequently, also methods of verification would have to be differentiated. In order to facilitate verification, it was found advisable to separate chemicals into categories using

[1] A binary chemical weapon is a device filled with two chemicals of low toxicity which mix and react when the device is delivered to the target, the reaction product being a supertoxic warfare agent, such as nerve gas.

toxicity criteria as a supplement to the general purpose criterion. The standards adopted for identifying such categories are as follows:

(*a*) $LCt_{50} = 2\,000$ mg min/m^3 for inhalation and/or $LD_{50} = 0.5$ mg/kg for subcutaneous injections; and

(*b*) $LCt_{50} = 20\,000$ mg min/m^3 for inhalation and/or $LD_{50} = 10$ mg/kg for subcutaneous injections.

LD_{50} and LCt_{50} indicate the amounts of toxic materials which are expected to kill 50 per cent of a large population of similar animals. LD is expressed in milligrams of toxic material per kilogram of body weight, while LCt is the concentration of the toxic material in the air, in milligrams per cubic metre, multiplied by the time of exposure in minutes. These values have meaning only when certain additional factors are specified, such as the species of animal to which they relate. This is one of the matters still under negotiation.

The use of additional criteria (for example, structural formulae, or applicability for chemical warfare) is also envisaged. In any event, the idea to compile a catalogue of *outlawed* chemicals, as put forward in the Disarmament Committee a few years ago, has apparently not been taken up by the two powers. Indeed, an exhaustive list of prohibited items is well-nigh impossible to draw up and keep up-to-date, while an incomplete list may be interpreted as permitting items not included therein. On the other hand, for reporting purposes, it may be necessary to establish lists of important chemicals in widespread *civilian* use, which could be diverted to chemical weapons purposes.

The two powers recognize the necessity for states to declare, "immediately after they become parties to the convention", the volumes of their stocks of means for chemical warfare and the means of production of chemical munitions and chemicals covered by the convention. Plans for destruction of declared stocks and for destruction or dismantling of means of production would also have to be declared. However, the specific contents of these declarations remain to be negotiated. Moreover, the term "means of production", as applied to the convention, has to be defined; declared reductions of the overall capacity of a country to produce chemical munitions and chemicals would be difficult to check, while monitoring the elimination of individual facilities previously engaged in chemical weapons production could be a simple operation.

The two sides agree that stocks of means for chemical warfare should be destroyed or diverted for permitted purposes within 10 years after a state becomes a party. Means of production should be shut down and eventually destroyed or dismantled; such destruction or dismantling should begin not later than 8 years, and should be completed not later than 10 years, after a state becomes a party. Such a long period of time envisaged for the implementation of the relevant provisions of the future convention is probably inevitable, in view of the size of the accumulated stocks and the technical problems involved in the destruction operations, as well as of the environmental considerations and safety requirements. The parties would probably begin with the elimination of obsolete stocks of munitions which they would have eliminated anyway, even without a treaty. Nevertheless, provided that it starts early and continues uninterruptedly, in coordinated stages, until completion, an extended process of weapons elimination may help to attenuate the acuteness of the otherwise insoluble problem of undeclared, hidden stocks, because it would allow time for a gradual build-up of confidence among parties. It can also reduce the military advantages a non-party might derive from remaining outside the convention, because the parties would retain a chemical warfare capability for quite a long time.

II. Verification of compliance

There is a consensus that the implementation of the obligations assumed by the parties should be "adequately" verified, and that verification should be carried out through national and international means.

As far as international means are concerned, an important arrangement agreed upon is to create a consultative committee open to all parties and having a permanent secretariat. The exact powers and responsibilities of this committee are to be determined, but it is already clear that it would serve for the exchange of data on chemicals and precursors for permitted purposes, according to agreed lists, as mentioned above. It could also be used to channel requests for information from parties suspected of violations, as well as requests for on-site investigation. Upon the request of any party, or of the UN Security Council, the consultative committee would take steps to establish the "actual state of affairs"; the party suspected of a breach may or may not agree to on-site investigation, but in the latter case it would have to provide appropriate explanations. To enable the consultative body to start its work immediately after the entry into force of the convention, a preparatory committee would be set up upon signature of the convention.

While the two sides agree that in the case of civilian chemical industry routine checks would be both impractical and undesirable, and that *optional* inspection, "by challenge", would be enough, they differ as regards the need for *mandatory* international on-site inspection in other cases. The USA considers the latter type of inspection indispensable to monitor the destruction of stockpiles of chemicals and weapons, and the moth-balling and eventual destruction of chemical weapons production and filling facilities, as well as to check the facilities for permitted production of chemicals which are "primarily useful for chemical weapons purposes".

National measures of verification are usually understood to include reconnaissance satellites or extra-territorial sensors. The parties would undertake not to impede this kind of verification by resorting to deliberate concealment. It should be noted, however, that such sophisticated technical means are a virtual monopoly of the great powers.

It is further assumed that, in order to ensure compliance with the convention, states would take legislative measures in accordance with their constitutional procedures, and that they would set up such national mechanisms as they may deem necessary to enforce the adopted laws and regulations. Possibilities for confidence-building measures to be taken before and after the entry into force of the convention are under consideration.

III. Entry into force and withdrawal

The conditions for entry into force of the convention have not yet been agreed upon. The Soviet Union insists that all the leading military states of the world, including all the permanent members of the UN Security Council, should become parties before the convention becomes effective. The question is, of course, important for the security of the prospective parties. Since France, the UK, the USA and the USSR (i.e., four out of the five permanent members of the Security Council), and other states militarily most significant, will be involved in the negotiating process as members of the CD, they are expected to adhere to a treaty commonly agreed upon. Thus, China's participation in the work of the CD and a positive attitude on its part towards chemical disarmament may help to resolve this problem.

The convention would include a withdrawal clause patterned after a provision appearing in other arms control agreements. This means that each party would have the right to denounce the convention once it had decided that extraordinary events had jeopardized its supreme interests. The parties could also turn to the UN Security Council with complaints against an offender and expect that some action would be taken in their favour by the Council, but in the prevailing political circumstances the threat of abrogation appears to be the primary means of enforcing a disarmament treaty.

IV. Summary and conclusions

The joint US–Soviet report marks progress in the bilateral negotiations for a CW convention, as regards both the scope and the verification of the envisaged prohibitions.

These are the most important agreements hitherto reached:

1. The CW ban will be comprehensive;

2. The substances banned will be defined on the basis of a general purpose criterion, supplemented chiefly by the criteria of toxicity;

3. Means for chemical warfare as well as means of their production will have to be declared immediately after a state becomes party to the convention, and destroyed or dismantled within 10 years;

4. An international consultative committee, with a permanent secretariat, will be set up for verification purposes; and

5. On-site investigation "by challenge" could be carried out in certain cases.

The most important questions which remain to be solved are as follows:

1. Whether the activities to be banned should explicitly include research and tests carried out with the intention of producing prohibited chemical agents; planning, organization and training for chemical warfare; as well as the use of chemical weapons.

As far as the use is concerned, a mere reference to, or confirmation of continuing validity of, the 1925 Geneva Protocol, which forbids asphyxiating, poisonous or other gases, would not be enough. The Protocol was concluded under the conditions of retention of chemical weapons, and in ratifying it, or in acceding to it, many countries formally reserved the right to employ these weapons against non-parties or in retaliation, and had made preparations for such employment. Moreover, there are still differences of opinion about the scope of the ban under the Geneva Protocol, in particular as regards the legality of the use in war of certain chemical agents.

A convention banning the possession of chemical weapons should expressly rule out their use in war, without qualification and under any circumstances. Alternatively, the parties could undertake to withdraw the reservations they had made in adhering to the Geneva Protocol. The convention should also make it clear, in accordance with the 1969 UN General Assembly resolution, that the prohibition of use applies to all chemical agents having direct toxic effects on man, animals or plants. A prohibition on planning and organization and especially on training of military personnel for chemical warfare could provide additional guarantees of non-use, while restrictions on military-oriented research and tests would reinforce the ban on development and production of chemical warfare agents and weapons.

2. Whether there should be any systematic on-site verification of compliance.

In abolishing an entire category of weapons which have already been used on a large scale in war, and which are capable of mass destruction comparable to that caused by nuclear weapons, the CW convention would become the first significant disarmament measure ever concluded. In view of the security aspects involved in such a radical step, the parties would need to assure themselves that the banned items had actually been abolished, and that new ones were not being manufactured. Unilateral, unchecked declarations by governments would not provide such an assurance, while self-verification exercised exclusively by nationally constituted bodies would not meet the required criteria of impartiality. Extra-territorial verification by national means is beyond the reach of the majority of states and, in any event, its usefulness is limited. International control is therefore irreplaceable. It could take different forms, the most important being on-site inspection, both sporadic and systematic. Sporadic inspection may be needed, for example, to investigate allegations of clandestine production of chemical warfare agents, or of their illicit use. But in the case of chemical weapons stockpiles, there is no reliable substitute for systematic on-site monitoring of the process of their destruction. There exists a body of evidence that on-site verification, whether sporadic or systematic, can be so devised as to rule out the disclosure of legitimate industrial, commerical or military secrets.

3. What kind of confidence-building measures could be taken before and after the entry into force of the convention?

In this connection, the following proposals have been made:

- official statements of national policies with respect to chemical weapons;
- gradual removal of secrecy surrounding chemical weapons through exchanges of information;
- visits of foreign technical experts to relevant chemical facilities;
- attendance of military exercises by foreign observers;
- international co-operation in the field of protection against toxic chemicals.

These measures are conceived mainly as voluntary acts on the parts of states. The purpose of the first four is to fill some inevitable gaps in the verification procedures which, as is generally admitted, cannot provide complete assurance. The purpose of the fifth is to spread knowledge about anti-chemical-warfare measures, so as to guard against a risk that chemical weapons might be used either by violaters of the 1925 Geneva Protocol and the envisaged convention, or by non-parties.

Since the signing of the Geneva Protocol in 1925, chemical and biological weapons might be used either by violators of the 1925 Geneva Protocol and public mind, were dealt with together in international negotiations aimed at their elimination. A split occurred in 1972, with the conclusion of a convention prohibiting only biological and toxin weapons. Nevertheless, it was then generally recognized that this convention represented a first step towards the achievement of agreement on effective measures also for the prohibition of chemical weapons, and the parties expressly committed themselves to continue negotiations in good faith with a view to reaching "early" such an agreement. It will be noted that by 1 September 1979, that is, almost five years after the entry into force of the BW Convention, only some 55 per cent of the UN membership were party to it (see appendix 3). This poor record of adherence may, in part, be attributed to the slowness of negotiations for chemical disarmament.

Although it has, so far, been negotiated only bilaterally, between the USA and the USSR, the CW convention is meant to be a generally acceptable multilateral treaty. It is, therefore, essential that at an appropriate stage the negotiations themselves become multilateral. This stage seems to have been reached with the submission of the 1979 US–Soviet report recording a convergence of views between the two most powerful chemical weapons states on a series of key issues described above. There is a need to proceed to multilateral negotiations before the two powers agree on a complete treaty text, because the solution of the outstanding questions may require an input by other nations concerned with chemical disarmament, especially those which also possess chemical weapons or are in a position to manufacture them. Bilateral or regional agreements for chemical arms control might usefully supplement a universally applicable multilateral treaty, but cannot replace it. For, as stated in the Final Document of the UN Special Session on Disarmament, the *complete* and effective prohibition of *all* chemical weapons and their destruction represent one of the most urgent measures of disarmament, and the conclusion of the convention to this end is one of the most urgent tasks of *multilateral* negotiations.

APPENDICES

Appendix 1

Joint USSR–United States report on progress in the bilateral negotiations on the prohibition of chemical weapons

During the Vienna meeting of the leaders of the United States and the USSR in June 1979, both sides affirmed the importance of a general, complete, and verifiable prohibition of chemical weapons and agreed to intensify their efforts to prepare an agreed joint proposal for presentation to the Committee on Disarmament. The USSR and United States delegations are guided by this provision at the 10th series of the bilateral negotiations, which began on 16 July 1979.

In the negotiations, the United States and USSR delegations take into account the fact that prohibition of chemical weapons is, as was stressed in the Final Document of the United Nations General Assembly Special Session on Disarmament, one of the most urgent and vital problems in the area of disarmament. They are also guided by the requirement that a convention on the prohibition of chemical weapons, as any other international agreement in the field of arms control and disarmament, should enhance rather than diminish the security of the parties.

The USSR and United States delegations, taking into consideration the interest expressed by many delegations in the Committee on Disarmament concerning the status of the bilateral negotiations on a prohibition of chemical weapons, present the following Joint Report:

1. The two sides believe that the scope of the prohibition should be determined on the basis of a general purpose criterion. Parties to the convention should assume the obligation never in any circumstances to develop, produce, stockpile, otherwise acquire or possess, or retain super-toxic lethal chemicals, other lethal or highly toxic chemicals or their precursors, with the exception of chemicals intended for permitted purposes of

such types and in such quantities as are appropriate to these purposes, as well as chemical munitions or other means of chemical warfare. Negotiations are continuing on several issues relating to the scope of prohibition.

2. Permitted purposes are understood to mean non-hostile purposes (industrial, research, medical, or other peaceful purposes, law-enforcement purposes, and purposes of development and testing of means of protection against chemical weapons), as well as military purposes not related to chemical warfare.

3. In order to facilitate verification, it would be appropriate to use, in addition to the general purpose criterion, toxicity criteria and certain other provisions.

4. Agreement has been reached on the following approximate values for the additional criteria of toxicity mentioned above:

(*a*) $LCt_{50} = 2\,000$ mg min/m^3 for inhalation and/or
$LD_{50} = 0.5$ mg/kg for subcutaneous injections;

(*b*) $LCt_{50} = 20\,000$ mg min/m^3 for inhalation and/or
$LD_{50} = 10$ mg/kg for subcutaneous injections.

On the basis of these criteria, it will be possible to separate chemicals into appropriate categories, to each of which the general purpose criterion would be applied.

5. Different degrees of prohibition and limitation as well as differentiated methods of verification would be applied on the basis of these toxicity criteria and certain other provisions. These issues continue to be subjects of negotiations.

6. Negotiations are also continuing on definition of terms and several other issues.

7. The two sides have agreed that parties to the convention should assume an obligation not to transfer to anyone, whether directly or indirectly, the means of chemical warfare, and not in any way to assist, encourage, or induce any State, group of States, or any organization to carry out activities which parties would undertake not to engage in pursuant to the convention.

8. The two sides have come to an understanding regarding the necessity for States to declare, immediately after they become parties to the convention, both the volumes of acquired stocks of means of chemical warfare and the means of production of chemical munitions and chemicals covered by the convention. Plans for destruction of declared stocks of chemical weapons should also be declared. These declarations should contain information on the volume and timetables for destruction of such stocks. Plans for destruction or dismantling of relevant means of production should also be declared. In the course of the bilateral negotiations, the two sides are continuing to make efforts to agree on the specific content of the declarations concerning stocks of means of chemical warfare and concerning means of production. In this connexion, the basic concept of means of production is also a subject that remains to be resolved.

9. Agreement has been reached that stocks of means for chemical warfare should be destroyed or diverted for permitted purposes within ten years after a State becomes a party. Means of production should be shut down and eventually destroyed or dismantled. The destruction or dismantling of means of production should begin not later than eight years, and should be completed not later than ten years, after a State becomes a party.

10. In this connexion, the United States and the USSR believe that a future convention should contain provisions in accordance with which parties would periodically exchange statements and notifications concerning: the progress of the destruction of stocks of means of chemical warfare or their diversion for permitted

purposes, the progress of the destruction or dismantling of means of production of chemical munitions and chemicals covered by the convention, and of the completion of these processes.

11. The USSR and the United States believe that the fulfilment of the obligations assumed under the future convention should be subject to the important requirement of adequate verification. They also believe that measures with respect to such verification should be based on a combination of national and international measures.

12. International verification measures should include the creation of a consultative committee. This committee could be convened as appropriate by the depositary of the convention, as well as upon request of any party.

13. The activities of the consultative committee in the interval between meetings should be carried out by a secretariat. The mandate of the secretariat is a subject of negotiations.

14. The participants should exchange, through the consultative committee or bilaterally, certain data on super-toxic lethal chemicals produced, acquired, accumulated, and used for permitted purposes, as well as on important lethal chemicals and the most important precursors used for permitted purposes. To this end, it is envisaged to compile lists of relevant chemicals and precursors. The two sides have reached a significant degree of mutual understanding in developing agreed approaches to the compilation of such lists. The scope of the data to be presented remains to be agreed.

15. Additional functions for the consultative committee remain under discussion.

16. In order to ensure the possibility of beginning the work of the consultative committee immediately after entry into force of the convention, the United States and the USSR believe it appropriate to begin the creation of a preparatory committee upon signature of the convention.

17. A convention should include provisions in accordance with which any party should have the right on a bilateral basis, or through the consultative committee, to request from another party with respect to which suspicions have arisen that it is acting in violation of obligations under the convention, relevant information on the actual state of affairs, as well as to request investigation of the actual state of affairs on site, providing appropriate reasons in support of the necessity of such an investigation.

18. A party may agree to such an on-site investigation or decide otherwise, providing appropriate explanations.

19. It should also be provided that any party could turn to the Security Council with a complaint which would include appropriate rationale. In case of suspicion regarding compliance with the convention, the consultative committee, upon request of any party, or of the Security Council of the United Nations, could also take steps to establish the actual state of affairs.

20. The question of other international verification measures remains unresolved.

21. National measures would include the use of national technical means of verification in a manner consistent with generally accepted principles of international law. In this connexion, parties should not impede, including through the use of deliberate concealment measures, the national technical means of other parties in carrying out the aforementioned verification functions.

22. The USSR and United States believe that a future convention should reflect the obligation of each party to take appropriate internal measures in accordance with its constitutional procedures to prohibit and prevent any activity contrary to the provisions of the convention anywhere under its jurisdiction or control.

23. Possibilities for confidence-building measures are being explored.

24. A future chemical weapons convention should include a withdrawal provision of the type included in other arms control and disarmament agreements.

25. The question of the conditions for entry into force of the convention remains unagreed.

26. The two sides believe that an effective prohibition of chemical weapons will require working out a large number of technical questions which would be dealt with in annexes to the convention and which are now being studied.

* * *

The United States and the Soviet Union note the great importance attached to the elaboration of a convention by the General Assembly of the United Nations and the Committee on Disarmament which manifested itself, in particular, in the identification of the question of the prohibition of chemical weapons as one of the priority items on the agenda adopted for the current session of the Committee on Disarmament. Both sides will exert their best efforts to complete the bilateral negotiations and present a joint initiative to the Committee on Disarmament on this most important and extremely complex problem as soon as possible.

Source: Committee on Disarmament document CD/48, 7 August 1979.

Appendix 2

Compilation of material on chemical warfare from the Conference of the Committee on Disarmament and the Committee on Disarmament working papers and statements, 1972–1979 (prepared by the Secretariat at the request of the Committee on Disarmament)

Figures in parentheses, thus (1), *refer to the list of references on pages 182–192.*

1. *Introduction*

In 1971, the CCD started negotiations on a ban on biological weapons separately from negotiations on chemical weapons, but with the understanding that the final objective remained the prohibition and elimination of chemical weapons as well. It was also agreed that toxins would be included in the ban, thus significantly broadening its scope.

The Convention on the Prohibition of the Development, Production and Stockpiling of Bacteriological (Biological) Toxin Weapons and on their Destruction was concluded by the CCD in 1971. It was commended by the General Assembly in resolution 2826 (XXVI) and entered into force on 26 March 1975.

In the Preamble of the Convention the States Parties recognized that an agreement on the prohibition of biological and toxin weapons represented a first possible step towards the achievement of agreement on effective measures also for the prohibition of the development, production and stockpiling of chemical weapons. Moreover, in Article IX of the Convention, each State Party affirmed the recognized objective of effective prohibition of chemical weapons and, to that end, undertook to continue negotiations in good faith with a view to reaching early agreement on effective measures for the prohibition of their development, production and stockpiling and for their destruction, and on appropriate measures concerning equipment and means of delivery specifically designed for the production or use of chemical agents for weapons purposes. Article VIII of the Convention also provided that nothing in the Convention should be interpreted as in any way limiting, or detracting from the obligations assumed by any State under the Protocol for the Prohibition of the Use in War of Asphyxiating, Poisonous and Other Gases, and on Bacteriological Methods of Warfare, signed at Geneva on 17 June 1925.

Beginning with its twenty-sixth session in 1971, and at its subsequent sessions the General Assembly adopted a number of resolutions* by which it requested the CCD and subsequently the CD to continue negotiations as a matter of high priority with a view to reaching early agreement on effective measures for the prohibition of the development, production and stockpiling of chemical weapons and for their destruction. It also invited all States that had not yet done so to accede to the 1925 Geneva Protocol and called for the strict observance by all States of the Objectives contained therein.

From 1972 to 1979 the CCD and the CD devoted intense efforts to the question of chemical weapons. It considered in detail all the main aspects of the question, including the scope of a ban, verification, complaints procedures and other related matters.

---

*Resolutions 2827A (XXVI), 2933 (XXVII), 3077 (XXVIII), 3256 (XXIX), 3465 (XXX), 31/65, 32/77, S-10/2, 33/59A and 33/71H.

The main points considered by the CCD and the CD in this context are referred to below under the relevant sections.

2. *Scope of ban*

With respect to the scope of a ban on chemical weapons, two main approaches have been discussed: a comprehensive ban in one step and a step-by-step approach.

2.1. *Comprehensive ban*

The following suggestions have been made in connexion with the question of a comprehensive ban:

Development, production, stockpiling, acquistion or retention of chemical warfare agents and of chemical weapons should be prohibited (1) as well as their use (2). These activities should also be prohibited with regard to munitions, equipment and means of delivery (3). Chemical warfare agents, chemical weapons, equipment and means of delivery should be destroyed or diverted to peaceful use (4). Various kinds of military activities, such as offensive military training, should be prohibited (5).

Parties to any convention banning chemical weapons, in addition to assuming the obligations as determined by the scope of the convention, should undertake not to transfer to any recipient whatsoever and not in any way assist any State or international organization to manufacture or acquire any of the agents, weapons, equipment or means of delivery to be banned under the convention (6).

A convention should not hinder measures for acquiring protection against chemical warfare (7), including assistance (8), prophylaxis and medical treatment (9).

In order not to encourage the development of chemical warfare agents, it should be prohibited to issue patents for chemical warfare agents and presently existing patents should be voided (10).

The convention should provide for adequate verification (11) or a system of guarantees to ensure that all parties are complying with the obligations (12).

Adequate verification should be based on a combination of national and international arrangements, including the creation of a consultative committee (13).

Upon signing or adhering to a convention (14), or when a convention enters into force (15), parties should declare their possession of chemical weapons, agents and production facilities (16).

Any such convention should be implemented in a manner designed to avoid hampering the economic or technological development in the field of peaceful chemical activities (17).

In a convention banning chemical weapons nothing should be interpreted as in any way limiting or detracting from the obligations under the Geneva Protocol of 1925 as well as under the Biological Weapons Convention (18).

The need for a further general position paper with a broad covering of the subject has been expressed (19).

2.2. *Step-by-step approach*

Various suggestions, as indicated below, have been made in connexion with this approach, on the assumption that a comprehensive ban could not be attained in one step. What these suggestions have in common is that they define initial, limited steps which would help achieve a comprehensive ban at a later stage (20). The view has been expressed that a partial approach introduces new technical elements and may stimulate military interest in sectors not covered by a partial agreement. Binding obligations to continue negotiations would then be needed (21).

One suggestion is that a moratorium would be declared on the development, production and stockpiling of the most lethal chemical warfare agents, pending agreement on the prohibition of such weapons (22).

According to another proposal, the first stage would consist of a ban on the development, production and stockpiling of supertoxic chemical warfare agents, together with the destruction of such agents (23).

Another formula envisages, as a first step, the conclusion of a convention dealing with the most dangerous, lethal means of chemical warfare (24).

The prohibition of all lethal chemical warfare agents, with or without a phased destruction of stockpiles, has also been envisaged (25).

Another proposal covers lethal chemical agents and other toxic chemical agents intended primarily to cause long-term physiological harm to human beings, with phased destruction of such agents (26).

Verifiability of production of chemical warfare agents (see 2.4.2 below) has been proposed as another possible criterion (27). This criterion might be applicable to the production of nerve gases, belonging to the so-called organophosphorus chemical compounds, which include many compounds having peaceful uses (28).

Some comments have been made regarding the delimitation of weapons, equipment and delivery systems to be banned (29). In one instance reference was made to munitions instead of weapons, with a view to covering binary chemical weapons (30).

With regard to activities to be prohibited, a first step might comprise a ban on production without destruction of stockpiles (31). On the other hand, stockpiles might be destroyed, while production facilities would be kept "moth-balled" (32).

A phased approach with regard to gradual destruction of stocks but within a comprehensive scope has also been suggested (33).

A comprehensive ban might be reached in steps by bringing under the ban, at appropriate times, items which had been left out (34). The scope of a convention should not be such as to be discriminatory against certain countries (35). The use of chemical weapons may be more probable in regional conflicts than in a major war (36). Regional agreements might prove to be useful supplements to a convention and could increase the prospect that the international community accept a chemical weapons ban (37).

Principles for delimiting the scope might be set out in a protocol to the convention or drafted and finalized outside the text of the basic agreement (38).

A method of delimiting the scope would be to list the chemical warfare agents to be covered by a convention (39). In this connexion it has been suggested that the warfare agents which the parties agreed to ban should be listed in an annex to the convention. As an alternative, one might list those agents which were to be exempted from a ban (40). One could also exempt for some time activities to be prohibited rather than agents (41).

The principle of delimitation would also apply with respect to the distinction between activities and agents related exclusively to warfare use, referred to as "single-purpose", and those which might also have a peaceful use, referred to as "dual-purpose" (42). A way of dealing with this problem might be to make a prohibition of single-purpose activities and agents unconditional, while the prohibition would be conditional with regard to the dual-purpose activities and agents (43). The so-called "purpose criterion" (see 2.4.1 below) could also be used in dealing with this problem (44).

2.3. *Definitions regarding scope*

2.3.1. *Activities*

With respect to the scope of activities which might be banned, three main categories of activities have been considered, namely, development, production and stockpiling, covering agents as well as weapons. Also planning, organization and training for offensive purposes have been mentioned.

In connexion with a phased agreement, the question has been raised whether such an agreement should initially encompass all activities affecting only certain CW agents, certain activities affecting all agents, or certain agents as well as certain activities (45).

Possible new elements of importance for the formulation of the scope of a convention may be identified in the process of the elaboration of the draft convention (46).

Reference has been made repeatedly to the fact that peaceful activities must not be hindered or interfered with (47).

With regard to development, production and stockpiling, it has been proposed that the general purpose criterion should apply (48). It has been suggested to examine the possibility that new types and new systems of chemical weapons be covered by an agreement on new types and new systems of weapons of mass destruction (49).

2.3.2. *Chemical weapons*

Chemical weapons have been described as combinations of the effective component—the chemical warfare agent—and the means and organizational structures for their military use (50).

Binary chemical weapons have been described as chemical weapons in which two less toxic chemical agents react to form a highly toxic agent on the way to the target (51).

Chemical weapons are considered to be weapons of mass destruction (52). It has been pointed out that there is a real danger that chemical weapons may be gradually assimilated and accepted as conventional weapons, if no agreement is reached to ban them (53).

It has also been suggested that "multi-purpose chemical weapons" causing physiological as well as mechanic and thermal effects should be treated as chemical weapons (54). The possible effect of chemical weapons on civilian populations and their sources of food and water make them detrimental to national and international security (55).

2.3.3. *Chemical warfare agents*

Chemical warfare agents are chemical substances which might be used in war because of their toxic properties (56). Effects on animals and plants should also be considered (57). Chemicals used in war for other purposes as, for instance, explosives, gun powder, fuel, smoke-generating chemicals, lubricators and napalm, etc., (58) have effects which are physical in their nature and do not belong to the category of chemical warfare agents.

Those chemicals which are precursors to the active agents in binary chemical weapons are in a special position (59). The purpose criterion might apply to them as well as to incapacitating agents (60).

A detailed definition of chemical warfare agents, including binary components, might be provided for in a protocol to a convention (61).

Several criteria have been suggested to describe the toxic properties of chemical warfare agents. They refer to the different toxic effects with regard to men, animals and plants (62). They are related to the various types of toxic effects (63) and penetration routes depending on the toxic agents involved (64). The degree of toxicity has been suggested as a criterion for determining the delimitation of single and dual purpose chemicals from each other, and from those with only peaceful use, as well as for delimiting super-toxic substances from less toxic ones (65).

Not only toxicity but also other properties have to be taken into account when evaluating a chemical as a possible chemical warfare agent (66). Relationship with respect to chemical structure has been mentioned as one delimitation criterion (67). One should also consider that the absence of protection and medical treatment facilities against chemical warfare may make less toxic agents suitable as chemical warfare agents in an attack against a country (68). Some of these criteria have been discussed (69) and their relationships for delimitation purposes have been analysed (70).

Toxins are already covered in the BW Convention, but since they have been characterized as lethal chemical substances (71), and in order to avoid ambiguities in interpretation, it has been suggested that they should be explicitly mentioned also in any future CW convention (72). Corresponding views have been expressed with respect to herbicides and defoliants about which it has been noted that certain restrictions apply to their use under the Enmod Convention and the new Protocols to the Geneva Conventions on Humanitarian Law in Armed Conflicts (73).

Recently, consideration has been given to some chemical agents with respect to their particular effects, such as those resulting from delivery of non-toxic or low-toxic agents to the target area, where they, either by reacting with components in the target or making it possible for components in the target to react with each other, result in some detrimental effects which directly or in the long run may cause harm to human beings (74).

2.4. *Delimitation criteria*

2.4.1. *Purpose and quantity criteria*

One important way to distinguish between activities and weapons (including agents) which are to be prohibited and those which are not to be prohibited, is to look for the underlying purpose, i.e. the general purpose criterion (75). Accordingly, all single-purpose activities and agents having use only for war should be unconditionally prohibited. The purpose criterion might also cover incapacitating agents as well as agents which may be developed in the future (76). Dual-purpose agents might be only conditionally prohibited, i.e. allowed as long as no warfare use was intended (77). The presence or absence of such intentions might be judged from the amounts of possible warfare agents and equipment involved. This quantity criterion is closely connected with the purpose criterion (78). With regard to the basis for justification of the quantities produced, which may vary considerably from thousands of tons to just a few kilograms a year (79), suggestions have been put forward aiming at national analysis of open-production statistics by the parties to a convention (80) (see 3.2.3.2 below), or reporting or declarations of production to some international body for further analyses (81).

2.4.2 *Verifiability criteria*

There exists an inter-relationship between attainable prohibitions and the potential of various approaches to verification (82).

One criterion for deciding whether production of a potential chemical warfare agent shall be banned or not is the verifiability of the production (83). The application of this criterion to at least the organophosphorus compounds has been considered to be possible (84), as these compounds have a relatively homogenous basis in the consumption of certain types of phosphorus and certain derivatives of that element. (Verification issues are dealt with in detail in section 3.2 below).

2.4.3 *Effect criteria*

The main property that makes a chemical substance a chemical warfare agent is its toxic effect. One definition of "toxic" that has been suggested is "poisonous in the sense of causing physiological injury to a human; this includes blistering, blindness and death" (85). The expression "long-term physiological harm to human beings" has also been used in this connexion (86). Accordingly, a system to determine toxicities of chemical substances was suggested early during the deliberations on chemical weapons (87). This system has been regarded as a possible tool for determining the scope of a ban with respect to chemical warfare agents (88). One might, for instance, differentiate super-toxic, single-purpose agents from less toxic, dual-purpose agents (89). Two different toxicity limits might be used to delimit such less toxic dual-purpose agents from super-toxic agents and from other chemical substances which cannot be used as warfare agents (90). A step-by-step approach might utilize the purpose criterion, supplemented by the toxicity criterion (91).

Delimitation might be facilitated by combining structure and toxicity criteria, at least with regard to super-toxic, organophosphorus compounds (92).

Some of the technical discussions in the CCD have been devoted to suggestions on actual numerical values of suitable toxicity limits (93). With regard to suitable limits for super-toxic agents, several figures have been suggested, which are all within the same order of magnitude (94). These differences, nevertheless, imply that potentially important chemical warfare agents may or may not be delimited as super-toxic agents, among them binary and multicomponent weapons (95). Methods to establish the limits as well as the degree of confidence that may be placed in the different approaches have been discussed (96). The need for evaluating and agreeing on standardized procedures for toxicity determination have been stressed (97).

Somewhat different methods must be used to determine the toxic effects—other than lethal—of harassing and incapacitating agents (98).

2.4.4 *Chemical structure*

The scientific system for describing the chemical structure of chemical compounds allows prediction of the structure even of compounds which may not yet have been synthesized. Thus it is theoretically conceivable to describe entire groups of toxic, chemically related compounds and suggest that such groups should be subject to prohibition under a convention. Therefore, one should, in theory, be able to cover in a convention even compounds similar to existing chemical warfare agents but not yet synthesized (99). Examples of groups of chemical agents which might be delimited in this way are the super-toxic organophosphorus compounds, which include the nerve agents, and also certain binary weapons components, mustard-type agents and arsines (100).

2.4.5 *Other properties*

Chemical substances may be very toxic, but, for a number of reasons, may be unsuitable for use in chemical weapons. By evaluating to which extent the properties of a chemical substance fulfil certain requirements, one could develop a method to determine whether a chemical substance can be considered to fall under the prohibitions of a CW ban (101). Such properties comprise, among others, shelf-life, volatility and explosion stability. By giving properties weighted numbers, a combination of such scales could provide an index or "evaluation number" that could help determine whether chemical substances can be classified as chemical warfare agents (102).

A way to establish whether known chemical agents fall under a ban is to list them (103). Combinations of criteria like toxicity and structural properties have been suggested in order to reduce the number of substances which might be necessary to list (104).

Lists, both of banned and exempted agents, might appear in one annex to a convention, implying that agents not mentioned in either of them would be covered by the purpose criterion of the convention which, in effect, would be comprehensive (105). Such lists could be reviewed and updated from time to time. This dynamic character of the scope of the convention should be the subject of closer analysis, with a view to determining whether this might provide for an effective step-by-step procedure towards a comprehensive agreement (106).

Using the (amended) single Convention on Narcotic Drugs of 1961 as a model other types of lists of CW agents could be constructed with a special view to the desirability of different types of verification. For this purpose the UNEP International Register of Particularly Toxic Compounds may be useful (107).

3. *Compliance*

With respect to the question of compliance, the following aspects have been considered.

3.1. *Confidence-building measures*

It has been suggested that confidence-building measures might occur both before and under a convention (108).

Countries should declare their CW policies and those possessing chemical weapons should declare their stocks (109). Parties could declare their possession of chemical weapons and production facilities either upon signing the treaty (110), when ratifying it (111), or upon its entry into force (112). With respect to requests for information the principle of equal security has been evoked (113). Other confidence-building measures, prior to the conclusion of a convention, might be to invite other countries to observe destruction of declared stocks (114) or to arrange technical exchange visits to selected facilities (115). Invitations to technical visits were discussed in the Committee on Disarmament and results from the visits have been described (116). Exchange of information on protection activities might also serve as a confidence-building measure (117). Partial agreements might themselves be confidence-building with respect to reaching a comprehensive ban (118). It has been pointed out that confidence-building and verification are different concepts that should be separated (119). Further consideration of confidence-building measures might be valuable (120).

3.2. *Verification measures*

3.2.1 *National verification measures*

One basic approach to the problem of verification is that the convention should to some degree be an expression of trust among countries (121). A nation's continued assurance that a convention is complied with should be based on utilizing national means of verification in combination with some international measures (122). This might imply setting up particular national verification organizations or control committees. Members of the committees might be representatives of governmental agencies, public organizations and experts. Their task should be to monitor that no violations against the convention take place within a country. A national verification organization might also exchange and analyse nationally and internationally available information (123). National means of extraterritorial control could utilize a combination of remote monitoring, including the use of satellites, and indirect monitoring by means of statistical data analysis (124). Such an organization should put forward suggestions regarding necessary national legislation for compliance with the treaty (125). One task might be to report information about national activities to an international verification agency (126). Countries which lack national technical facilities for establishing verification measures on their own might be put at a disadvantage if a convention only provided for national means of verification (127). In order to work out standardized programmes for national verification agencies, it might be useful to arrange international expert conferences or establish basic principles internationally (128). The possibility of using supervision procedures in addition to national means may be investigated (129).

3.2.2 *International verification measures*

The view has been expressed that a country in possession of a chemical weapons capability gives up a significant military option if it becomes party to an agreement banning such weapons (130) and that it would renounce the possibility of "retaliation in kind" if it were attacked with chemical weapons. It would also lose the deterrent effect such a capability might have (131). For the sake of its own security, a country may wish to include in the agreement verification provisions designed to prevent other parties to the convention from secretly preparing or maintaining a chemical weapons capability (132). Such provisions have been suggested specifically to comprise international verification measures, including on-site inspections (133). Regularity of inspections would enhance confidence-building (134). These measures should be sufficiently effective to actually deter parties from possible violations (135). The verification measures need not be 100 per cent effective (136). States armed forces should be exempted from international verification (137). In addition to its deterring effects, an international verification system would provide continuous reassurance to the parties that no violations were occurring (138). A country having been subject to on-site verification can share its experience with other countries (139). International verification measures can be supplemented by national ones (140). Concerning the degree of intrusiveness of methods, on-site inspection has been considered to be too intrusive (141). However, different degrees of intrusiveness can apply also to on-site verification activities (142). Different international verification methods, applied together, will reinforce each other (143). Non-intrusive international verification measures may observe certain activities and involve analysis of the observations, in order to obtain indications of possible violations as a basis for further verification and complaints procedures (144).

3.2.2.1 *Organizational aspects*

International verification activities could be conceived of as taking place on a voluntary basis in co-operation with national control committees (145). Consultative committees could also be established from among the parties themselves as a result of a formal agreement (146).

The agreed type of information prepared by national verification committees, or otherwise available, could be circulated, studied and analysed for consistency either by particular expert groups called upon by the Parties to the Convention (147), by a consultative body or committee (148), or by a verification agency (149). The Secretariat of the United Nations might be assisted by experts in considering verification problems (150). A comprehensive scheme would comprise an international body authorized to carry out verification, when so requested by a party or on its own (151). Such a body could also be entrusted to carry out on-site inspections (152). Preparation of technical material, for use in the implementation of a chemical weapons ban, for instance in the form of an "analytical handbook" for chemical analyses, is already being undertaken at this stage (153).

With regard to international verification agencies, different suggestions regarding their names and functions have been put forward (154). Some suggestions also consider the specific need for international verification of a chemical ban together with the question of international verification of a disarmament agreement in general (155).

Existing international organizations with technical resources, like WHO and UNEP, might suitably take on certain monitoring activities to ensure compliance with a treaty (156). They could, for instance, collect technical information on properties of chemical substances and methods of chemical analyses and also provide experts (157).

Costs and manpower requirements for any international verification activities and organizations should be kept as low as possible (158). There is a risk that international verification measures may lead to unwanted and illegal disclosures of a nation's military, technical and industrial secrets to other nations (159) and some measures to avoid such risks have been suggested (160). Information on chemical agents listed in annexed lists to a convention could contribute to verification measures (161).

3.2.3 *Verification for specific activities*

3.2.3.1 *Development**

Much development in the field of chemical weapons has originated from research and development for peaceful purpose, sometimes far in advance of the actual weapons development (162). One way of getting an early indication on potential chemical weapons applications is a systematic, computerized search of internationally available scientific and technical literature on a routine basis (163). Open reporting and internationalization of information has been called for (164). Suggestions on voluntary exchange of information have taken into account results of scientific research and developments for peaceful purposes (165).

One activity that is related to the development of chemical weapons is field testing. Field tests might be detected and monitored by so-called remote sensing methods (166). What is usually meant by remote sensing is the use of analytical equipment that can provide information on phenomena at a distance from the analysing equipment or the

* Some of the methods mentioned in this section might be applicable also to monitoring air, earth and water around production and destruction facilities and be considered also in the next three sections.

observer (167). The possibility of detecting field tests with nerve agents by means of satellite based spectrophotometric instruments (infra-red sensitive) have been analysed (168). An analysis has also been carried out regarding the capacity of similar instruments, as well as of others based on different principles, on the assumption that such instruments would be placed on earth but outside the borders of countries to be monitored. Geographical and meteorological conditions are found to influence all the methods discussed (169).

3.2.3.2 *Production*

Verification of non-production of chemical weapons presents the basic difficulty of proving the negative (170). Control measures can be carried out in production facilities resulting in assurance of non-production of chemical warfare agents, without disclosure of production secrets. Such measures could include regular on-site inspections, arranged by an international control agency (171).

Verification problems relating to the production of chemical agents, munition and equipment for delivery, and of facilities for filling the munition have been considered (172). Parties to a convention should declare means of production of chemical munitions and chemicals covered by the convention (173). The difficulties of obtaining access to military installations for verification purposes have been mentioned (174). Confirmation by some method that production for chemical weapons does not occur may be relevant only with respect to that particular verification activity and not necessarily prove that a violation had not occurred (175). Even if verification methods by means of which violations of a production ban can be devised, their practical application would be difficult due to the magnitude and diversity of the chemical industry (176).

Providing information on pertinent production activities for analysis by concerned parties constitutes one way of providing material for verification activities (177). Historically known chemical warfare agents might be continually listed as a basis for information on chemical production facilities (178). Reference has been made to the use of production statistics, transportation data, etc., as a means to follow the production of relevant chemical substances in different countries (179). As an illustrative example of such an accounting method, a verification system relating to the production of organophosphorus compounds, to which the nerve agents belong, has been worked out in some detail and a number of limitations, including possible evasions, have been described. That system envisages that both national and international organizations would operate. The system requires that information be available from independent sources. Verification personnel should be allowed to visit production sites and would require basic information from different production levels to check the overall balance of the system, when warranted by analyses of statistical material (180). Verification measures should encompass also other chemical warfare agents than those of organophosphorus origin (181).

With regard to proposals for verification of a ban on production of dual-purpose agents for warfare use, it has been indicated that the production of dangerous chemical agents is being increasingly brought under both national and international regulations (182).

Several methods to control the actual production of chemical warfare agents have been suggested. On-site inspection has been requested for verifying that proscribed chemicals are not produced in facilities for similar substances (183). It has been debated whether photographing from satellites or airplanes, or mere ocular observation from the outside of a production plant might provide useful information (184). Remote sensing methods for monitoring outlets and surroundings of chemical plants are conceivable under certain conditions (185). Highly sensitive chemical analysis of actual samples from such areas is also conceivable (186). Attempts may be made to detect those products that may develop when CW agents, as well as other chemical substances involved in their production, leak out into the surroundings. Also, the presence of certain binary weapons precursors could be demonstrated (187). Also, it may be possible to analyse and follow chemical traces in the environment for some time after a release, or in connexion with an alleged use (188).

One prerequisite for resorting to such methods is either to get access to the place from which the material to be analysed can be obtained, or to ensure that unmanned analytical equipment can function undisturbed at the place. The possibility of sampling of material by means of unmanned "black boxes" for actual chemical analysis *in situ* or elsewhere, as well as the usefulness of these methods, should be investigated (189).

In this connexion, several technical devices characterized as "on-site but non-intrusive" have been presented with the aim of ensuring that, after an agreement has been reached, relevant equipment or areas are not tampered with. These might include production facilities which, in accordance with the agreement, would cease production and remain "moth-balled" without on-site inspection—a technique that has also been developed to safeguard nuclear facilities. Monitoring of compliance of that agreement might then take place by means of a country's "own national means of verification" (190). Non-intrusive verification methods of "moth-balled" production facilities could not substitute for destruction or conversion to civilian uses of the facilities (191). Absence of safety measures in a production plant may be a sign that no CW agents are produced there, even if in some instances production of substances with low acute, but high chronic toxicity exist (192).

Possibilities for "familiarization exchange" of information, as discussed in connexion with other arms control treaties, have been suggested (193).

It has been suggested that by utilizing a proposed system for delimitation of potential chemical warfare agents by means of a combination of evaluations relating to the different properties of a substance (194), one might find a means of directing the activities of a verification agency with regard to development or production of chemical substances for potential warfare use (195).

It is considered that verification of production of chemical weapons must reflect the fact that this production has more in common with biological weapons than nuclear weapons (196).

3.2.3.3 *Stockpiling*

Discussion on this subject has dealt with munitions and bulk storage of chemical agents (197). Reference has been made to the problem of how to verify that chemical munition is not stored together with ordinary munition (198). Remote monitoring of munition transports may be the only way of finding secret stockpiles (199). A comprehensive ban might facilitate verification of munition stockpiles, since no chemical munitions at all would be allowed (200). The difficulty of finding hidden stockpiles has been mentioned (201). Declaration of stocks before the conclusion of a convention, or upon its entry into force, could inspire confidence among parties concerned (202), and facilitate planning of a destruction programme (203). A convention should contain provisions for declaration of stockpiles (204).

Monitoring the state of known stockpiles by air or satellite reconnaissance seems to have only limited possibilities (205). On-site visits to known stockpiles may confirm the nature of the stockpile, if admittance to the storage facilities is allowed. Possibilities would exist to find out whether stockpiles contain chemical weapons and what type of agent they contain, if measurements were allowed immediately outside the actual site (206). When stored over a long time, munition and bulk storages may begin to leak and deteriorate, necessitating adequate arrangements for taking care of such situations. Such measures may or may not be observable depending on whether particular precautions had been taken (207).

3.2.3.4 *Destruction of stockpiles*

Various technical aspects of destruction of stockpiles and the possible means of verifying such destruction have been discussed in a number of working papers (208). It has been suggested that the possibility of using additional supervision procedures for verifying destruction of stockpiles might be discussed (209).

It is conceivable that also undeclared stocks may be destroyed: for instance, in the case of the need to destroy "obsolete" munition (210). The possibility of verifying such destruction is, however, related to the problem of finding the stocks in the first place (211). The information contained in some working papers indicates that destruction of stockpiles is a protracted process involving, *inter alia*, hazards for the environment, and that rigorous procedures must be followed in carrying out such operations, which under some conditions might be observed (212). The consequences of a possible long destruction period on the formulation of a convention have to be taken into consideration (213).

Monitoring of destruction of stockpiles should account for the particular agent that is destroyed, and the quantity and quality of the destroyed agent, considering also weight and volume of other components in the stockpiles (214).

If on-site access to a stockpile destruction is not permitted, the activity cannot be verified by any presently known methods (215). Extraterritorial monitoring may be of some use (216).

Different suggestions have been given to find as non-intrusive verification methods for stockpile destruction as possible. One suggestion is that a country may choose a destruction site where on-site access would be acceptable (217). The destruction could be checked *in situ* by observers, as distinguished from inspectors (218). In this connexion, the possibility has been mentioned for the "familiarization" of other parties with the site of an activity through information exchange, a technique which has also been discussed in connexion with other arms control measures (219). Observation of destruction of stockpiles would not need to be looked upon as a recurrent on-site inspection measure, since a stockpile can be destroyed or converted to peaceful uses, and thus inspected once only (220).

The degree of disclosure related to the technical verification methods referred to above ranges from total disclosure of the destruction process of the agent, and the quantity being destroyed, to only an assurance that some toxic substance is being destroyed or converted into a less toxic one (221).

Ways of evading effective verification have been described (222).

The verification process should not result in unwarranted disclosure of military information leading to proliferation of chemical weapons or in disclosure of industrial secrets (223).

In connexion with suggestions for arranging technical exchange visits during ongoing negotiations, the usefulness of visiting also working facilities for destruction of chemical weapons has been mentioned (224). International co-operation in this field might be useful (225).

3.2.3.5 *Other military activities*

A chemical warfare capacity comprises not only development, production and stockpiling of chemical agents and weapons but also other activities like planning and training of personnel (226).

A comprehensive ban on chemical weapons might, perhaps, lead to observable changes in military doctrine, training, organization and equipment, and thus serve verification purposes (227).

The difficulty of distinguishing between defensive and offensive measures has been noted (228).

A convention should cover adequately the situation when a country has on its territory chemical weapons belonging to, and under the control of, another country (229).

As previously mentioned, it has been suggested that protective measures against chemical warfare should not be banned (230). In this connexion, it has been suggested that international co-operation regarding protection against chemical warfare might take the form of regular meetings of experts, or exchange of information, especially relating to organophosphorus poisoning, therapy and prophylaxis (231). It has also been proposed that a convention should ensure that support and assistance be rendered to a country victim of a chemical weapon attack (232).

3.3. *Complaints and clarification procedures*

A complaints procedure must be based on a number of interrelated measures to be appropriately included in any CW agreement (233).

The procedure for submitting a complaint to the Security Council under the provisions of the Charter could be spelled out in a convention (234). Due to the political nature of the decisions of the Council, it might be desirable to resort to international investigation and fact-finding procedures before a complaint is lodged with the Security Council (235).

Parties to a convention could undertake to consult with each other (236). Consultations could also take place within the multilateral framework of a consultative body (237), a consultative committee (238) or a verification organization (239).

Consultations might be arranged so that requests for clarification do not need to appear as formal complaints or allegations of violations with ensuing political difficulties (240).

A party to a convention which wanted to allay suspicions or respond to the provisions of the convention might take initiatives for verification by invitation (241). Verification might be an obligation under a convention providing for co-operation by a party when challenged by other parties (242). If such challenges were met with negative replies, they could lead to suspicion (243).

When consultations fail or if actual complaints are filed with a competent body, further measures might be taken requiring additional information, fact-finding investigations or inspections (244). With regard to fact-finding measures being taken as a result of complaints, adequate expertise for assistance (e.g. for chemical analyses and toxicity determinations) should be available either within the competent body itself or through experts or expert groups (245) available nationally or in international organizations (246).

It has been emphasized that possible conclusions by an international organization with respect to results from technical analyses and fact-finding should be expressed in language that can be easily grasped by agencies and personnel in developing countries and accordingly be useful when making complaints accompanied by evidence (247).

4. *Other provisions*

4.1. *Review conferences and amendments*

Review conferences should be held periodically (248). Review conferences should serve to ensure that the preamble and the provisions of the convention are complied with (249). Review conferences should take into account new technological and scientific developments relevant to the convention (250).

Parties should have a right to suggest amendments to the convention. Various systems for accepting such amendments have been envisaged (251).

4.2. *Technical assistance and use of disarmament savings*

Scientific and technical development in the field of chemistry should benefit peaceful activities and to this end, exchange of information and equipment for peaceful purposes should be facilitated (252).

The principle that a substantive part of savings derived from disarmament measures should be used to promote economic and social development, particularly in developing countries, must be recognized (253).

4.3. *Duration and withdrawal*

The convention could be either of unlimited (254) or limited duration (255).

When the supreme interests of States are threatened, parties should be able to withdraw from the treaty after prior notification (256).

4.4. *Adherence, entry into force, depositary agent*

Different provisions for signature, ratification and entry into force of the convention have been suggested (257). Governments could act as depositaries (258) or, as a result of discussions within the framework of the United Nations, another depositary could be designated (259).

4.5. *Protocols and annexes*

Some provisions and procedures of an agreement might be contained in protocols, annexes or "agreed interpretations". They might include principles for delimiting different kinds of chemical warfare agents (260), definitions, list of agents, reporting procedures (261), and stipulations for a possible consultative committee (262), or verification organization (263).

List of references

1. Socialist draft convention CCD/361, Article 1; Non-aligned working paper CCD/400, paragraph 7; Japan, draft convention CCD/420, Article I; United Kingdom, draft convention CCD/512, Article I; USSR, CCD/PV.788, p. 6; United States of America, CCD/PV.802, p. 20; Venezuela, CD/PV.29, p. 8.

2. United Kingdom, draft convention CCD/512, Article I; United Kingdom, PV.720, p. 10.
3. Socialist draft convention CCD/361, Article I; Non-aligned working paper CCD/400, paragraph 7; Japan, draft convention CCD/420, Article I; United Kingdom, draft convention CCD/512, Article I.
4. Socialist draft convention CCD/361, Article II; Non-aligned working paper CCD/400, paragraph 7; Japan, draft convention CCD/420, Article II:1; United Kingdom draft convention CCD/512, Articles III, VII.
5. Sweden, PV.499, pp. 8, 10; Sweden, CCD/322, p. 1; Yugoslavia, CCD/377, p. 2; Sweden, PV.622, p. 11; United States of America, PV.740, p. 24; Sweden, CCD/PV.764, p. 17; Sweden, CD/PV.2, p. 50; Venezuela, CD/PV.29, p. 8.
6. United States of America, CCD/360, p. 13; Socialist draft convention CCD/361, Article III; Non-aligned working paper CCD/400, paragraph 3; Japan, draft convention CCD/420, Article III; Yugoslavia, PV.714, p. 35; United Kingdom, draft convention CCD/512, Article VI.
7. Japan, draft convention CCD/420, Article I-a; United Kingdom, draft convention CCD/512, Article I-a.
8. Socialist draft convention CCD/361, Article VII; Non-aligned working paper CCD/400, paragraph 6; Japan, draft convention CCD/420, Article XI.
9. United Kingdom, draft convention CCD/512, Article XII-1.
10. USSR, PV.583, p. 13; Socialist working paper, CCD/403, p. 3.
11. Japan, PV.559, p. 11; Non-aligned working paper CCD/400, paragraph 9; United States of America, PV.613, p. 13.
12. Non-aligned working paper CCD/400, paragraph 11; USSR, PV.608, p. 15.
13. USSR, CCD/PV.788, p. 7; United States of America, CCD/PV.802, p. 20; Venezuela, CD/PV.29, p. 9.
14. Sweden, PV. 569. pp. 24–25; United Kingdom, draft convention CCD/512, Article II.
15. Netherlands, PV.560, p. 7; Non-aligned working paper CCD/400, paragraph III, A.a.
16. USSR, CCD/PV.788, p. 7; United States of America, CCD/PV.802, p. 20.
17. Socialist draft convention CCD/361, Article IX; Non-aligned working paper CCD/400, paragraphs 2, 4; Japan, draft convention CCD/420, Article XII; United Kingdom, draft convention CCD/512, Article XII.
18. Mongolia, PV. 552, p. 22; Socialist draft convention CCD/361, Article VIII; Egypt, PV.555, p. 11; Non-aligned working paper CCD/400, paragraph 1; Japan, draft convention CCD/420, Article XII; Nigeria, PV. 638, pp. 17–18; Japan, PV.661, p. 6; United Kingdom, draft convention CCD/512, Article XI; Venezuela, CD/PV.29, p. 8.
19. Italy, CD/5; Netherlands, CD/PV.8, p. 22; Netherland, CD/6.
20. United States of America, CCD/360, pp. 5, 8; United States of America, PV.551, p. 21; Japan, PV.631, pp. 7–10; Japan, draft convention CCD/420, Article IV; Australia, CD/PV.2, p. 37; Canada, CCD/PV.740, p. 9; Venezuela, CD/PV.29, p. 8.
21. Romania, CCD/PV.743, p. 8.
22. Mexico, CCD/346; Mexico, PV.545, p. 31.
23. Japan, PV.631, p. 9.
24. USSR, PV.642, p. 16; United States of America, PV.643, p. 17; Bulgaria, CCD/PV.731, p. 23.

25. United States of America, PV.687, p. 20; United States of America, PV.702, p. 8.
26. United Kingdom, draft convention CCD/512, Article I-a.
27. Canada, PV.496, p. 50; United States of America, CCD/360, p. 7; United Kingdom, PV.557, p. 6; United Kingdom, PV.575, p. 8; Japan, PV.588, p. 11; Japan, draft convention CCD/420, Article IV; Japan, PV.631, p. 8; United States of America, PV.702, p. 8.
28. Japan, CCD/430, p. 1.
29. United States of America, CCD/360, p. 6; Netherlands, PV.560, p. 8; Sweden, PV.622, p. 12.
30. United Kingdom, draft convention CCD/512, Article I-b; United Kingdom, PV.720, p. 10.
31. United States of America CCD/360, p. 8; Japan, CCD/413, p. 3; Canada, PV.643, pp. 19–20; United Kingdom, draft convention CCD/512, Articles III-d, VII; United Kingdom, PV.720, p. 10.
32. United States of America, CCD/360, p. 5; United Kingdom, PV.575, p. 11; Brazil, PV.579, p. 8.
33. Canada, PV.643, pp. 19–20; United States of America, PV.702, p. 11; Iran, PV.717, pp. 9–10; United Kingdom, draft convention CCD/512, Article VII; United Kingdom, PV.720, p. 9.
34. United States of America, CCD/360, p. 6; United States of America, PV. 537, p. 21; Japan, PV.631, pp. 8–9; Canada, PV.643, p. 19; United States of America, PV.702, p. 8.
35. Brazil, PV.597, p. 8; Netherlands, PV.560, p. 9; Sweden, PV.569, p. 28; Non-aligned working paper CCD/400, paragraph 9; Poland, PV.611, p. 9; United States of America, PV.702, p. 8.
36. Netherlands, CCD/PV. 741, p. 17.
37. Canada, CD/PV.23, p. 9.
38. USSR, PV.567, p. 18; Sweden, PV.569, p. 26; Non-aligned working paper CCD/400, paragraph 10; Poland, PV.611, p. 9; United Kingdom, CCD/PV.741, p. 31.
39. Italy, CCD/335, p. 1; United States of America, CCD/365, p. 5; Sweden, CCD/372, p. 7; Japan, CCD/430, p. 1; United States of America, CCD/499, p. 4.
40. Japan, draft convention CCD/420, Article IV, Annex I-A and B.
41. Canada, CCD/PV.740, p. 9.
42. United States of America, CCD/360, p. 1.
43. Sweden, PV.457, p. 19; Sweden, CCD/372, p. 2; Sweden, PV.569, p. 26; Canada, CCD/414, p. 5.
44. Socialist draft convention CCD/361, Article I; United States of America, CCD/360, p. 3; USSR, PV.567, p. 15; Japan, draft convention CCD/420, Article I.
45. United States of America, PV.702, p. 7.
46. Group of 21, CD/11, p. 3; Sweden, CD/PV.29, p. 32; Nigeria, CD/PV.31, p. 39.
47. Socialist draft convention CCD/361, Article IX-2; Non-aligned working paper CCD/4, paragraph 2; Japan, draft convention CCD/420, Article XIV-2; United Kingdom, draft convention CCD/512, Article XII-2.
48. Socialist draft convention CCD/361, Article I; Japan, draft convention CCD/420, Article I-a; United Kingdom, draft convention CCD/512, Article I-a; USSR, CCD/PV.788, p. 6; United States of America, CCD/PV.802, p. 20.
49. USSR, CCD/PV.736, p. 31.

50. USSR, PV.567, p. 18; Czechoslovakia, CCD/508, p. 1.
51. United States of America, CCD/360, p. 2; Sweden, PV.622, p. 10; Canada, CCD/414, p. 2; Yugoslavia, CCD/504, pp. 1–2.
52. Mongolia, PV.552, p. 22; Romania, PV.608, p. 21; Sweden, PV.721, p. 19.
53. Iran, PV.678, p. 14.
54. Yugoslavia, CCD/502, p. 2.
55. Nigeria, CD/PV.31, p. 40.
56. USSR, PV.567, p. 18; Yugoslavia, PV.569, p. 12; Sweden, CCD/427, p. 2.
57. Yugoslavia, CCD/PV.742, p. 8.
58. United Kingdom, PV.557, p. 7; Sweden, PV.635, p. 9; Yugoslavia, CCD/505, p. 2.
59. United States of America, CCD/360, pp. 2, 4; United States of America, CCD/365, p. 2; Sweden, CCD/427, p. 3.
60. United States of America, PV.360, pp. 3–4; USSR, PV.567, p. 15; Finland, CCD/381, p. 3; Sweden, PV.622, p. 1; Sweden, PV.652, p. 8; Iran, PV.678, p. 17; Yugoslavia, CCD/504, p. 2; United States of America CCD/PV.740, p. 24.
61. United Kingdom, CCD/PV.741, p. 31.
62. Sweden, PV.499, p. 8; Yugoslavia, CCD/375, pp. 1–2; Sweden, PV.569, p. 28; Yugoslavia, PV.569, p. 12; Nigeria, PV.638, pp. 17–18; Sweden, CCD/427, p. 2; Yugoslavia, CCD/505, p. 2; United Kingdom, draft convention CCD/512, Article I-a.
63. Canada, CCD/414, p. 5; United States of America, CCD/435, pp. 1–6; Canada, CCD/473, p. 4–5.
64. Canada, CCD/387, p. 7; United States of America, CCD/435, pp. 2–3; Canada, CCD/473, p. 1.
65. United States of America, CCD/499, p. 2.
66. Sweden, CCD/372, p. 3; Federal Republic of Germany, CCD/458, pp. 1, 3; United States of America, CCD/499, p. 1; Venezuela, CD/PV.29, p. 9.
67. Netherlands, CCD/320, p. 2; United States of America, CCD/360, p. 3; United States of America, CCD/365, p. 2; Japan, CCD/374, p. 4; Netherlands, CCD/383; Canada, CCD/414, p. 2.
68. Poland, PV.551, p. 26; Egypt, PV.555, p. 12; Yugoslavia, CCD/375, pp. 3–4; USSR, PV.567, p. 16; Yugoslavia, PV.569, p. 11; Non-aligned working paper CCD/400, paragraph 8; United States of America, PV.687, p. 20; Yugoslavia, CCD/503, p. 1; Czechoslovakia, CCD/508, p. 2; Poland, PV.611, p. 9; Czechoslovakia, CCD/PV.742, p. 18.
69. United States of America, CCD/499.
70. Sweden, CCD/461, p. 1.
71. USSR, PV.553, p. 30.
72. Sweden, PV.697, p. 21.
73. Netherlands, CCD/PV.758, p. 23.
74. Czechoslovakia, CCD/508, pp. 2–4; USSR, CCD/514, Annex, p. 1, paragraph 2a.
75. United States of America, CCD/360, p. 3, paragraph 134; United States of America, CCD/365, p. 4; USSR, PV.567, pp. 15–17; Hungary, PV.577; p. 9; USSR, CCD/PV.788, p. 6; United States of America, CCD/PV.802, p. 20.
76. Finland, CCD/381, p. 3; Mongolia, PV.616, p. 7; Sweden, CCD/461, pp. 3, 7–8; United States of America, CCD/331, p. 3.
77. Sweden, PV.457, paragraph 33; Sweden, PV.499, p. 5, paragraph 3, p. 9, paragraphs 19–20; United States of America, CCD/365, pp. 1–5; USSR, PV.567, p. 17; Argentina, PV.578, p. 7; Sweden, CCD/PV.764, p. 15.

78. USSR, PV.567, p. 16; Egypt, PV.572, p. 23; Sweden, CCD/461, p. 2.
79. Sweden, PV.499, p. 9, paragraph 20; Netherlands, CCD/PV.758, p. 24.
80. Socialist working paper CCD/403, p. 2; USSR, PV.714, p. 30.
81. Japan, draft convention CCD/420, Articles V, VI; United Kingdom, draft convention CCD/512, Articles II-1, c-d, VIII-b.
82. United States of America, PV.545, p. 14; United States of America, PV.551, p. 20.
83. Canada, PV.496, p. 50; United States of America, CCD/360, p. 7; United Kingdom, PV.557, p. 6; United Kingdom, PV.575, p. 8; Japan, PV.588, p. 11; Japan, draft convention CCD/420, Article IV; Japan, PV.631, p. 8; United States of America, PV.702, p. 8.
84. Japan, CCD/430, p. 1.
85. Canada, CCD/414, p. 1.
86. United Kingdom, draft convention CCD/512, Article I-a.
87. Japan, CCD/301, p. 1; Yugoslavia, CCD/375, p. 2.
88. United States of America, CCD/360, p. 2; USSR, PV.714, p. 28.
89. United States of America, CCD/360, p. 2; Sweden, CCD/372, p. 5.
90. Canada, CCD/414; Canada, CCD/473; Canada, PV.685, p. 15; United States of America, PV.702, pp. 9–10; USSR, CCD/PV.788, p. 7; United States of America, CCD/PV.802, p. 20.
91. USSR, PV.714, p. 28; USSR, PV.727, p. 23.
92. Japan, CCD/374, p. 4; Netherlands, CCD/383, p. 9; United States of America, PV.702, p. 10.
93. United States of America, CCD/499, pp. 2–3.
94. Japan, CCD/301; Netherlands, CCD/320; Sweden, CCD/322; Sweden, CCD/372; Japan, CCD/374, pp. 1–2; Canada, CCD/414; Japan, CCD/430; United States of America, CCD/435; Canada, CCD/473; Japan, CCD/515.
95. Japan, CCD/301; Netherlands, CCD/320; Sweden, CCD/322; Canada, CCD/414; Hungary, CCD/537/Rev.1; Hungary, CCD/PV.758, p. 20.
96. Japan, CCD/301; Netherlands, CCD/320; Sweden, CCD/322; Sweden, CCD/372; Japan, CCD/474, pp. 1–2; Canada, CCD/414; Japan, CCD/430; United States of America, CCD/435; Canada, CCD/473; Czechoslovakia, CCD/508; Japan, CCD/515; Hungary, CCD/PV.537/Rev.1; Hungary, CCD/PV.758, p. 19.
97. United States of America, CCD/365; Japan, CCD/374; Yugoslavia, CCD/375; Canada, CCD/387; Japan, CCD/430; United States of America, CCD/435; Canada, CCD/473; Japan, CCD/515.
98. Canada, CCD/433; Canada, CCD/473, p. 9; Sweden, CCD/PV.764, p. 15; United States of America, CCD/531.
99. Netherlands, CCD/320, p. 2; United States of America, CCD/360, p. 3; United States of America, CCD/365, p. 2; Japan, CCD/374, p. 4; Netherlands, CCD/383, p. 3; Canada, CCD/414, p. 2; Hungary, CCD/PV.753, p. 19.
100. United States of America, CCD/365, p. 3; Canada, CCD/414, p. 2; United States of America, CCD/497, p. 3.
101. Sweden, CCD/372, p. 2; Federal Republic of Germany, CCD/458, pp. 1,3; Federal Republic of Germany, PV.674, p. 6.
102. Federal Republic of Germany, CCD/458, p. 3; Federal Republic of Germany, PV.767, p. 6.

103. Italy, CCD/335, p. 1; United States of America, CCD/365, p. 5; Sweden, CCD/372, p. 7; Japan, CCD/430, p. 1; United States of America, CCD/499, p. 4; Australia, CD/PV.2, p. 37; Japan, CCD/529.
104. Japan, CCD/374, p. 3.
105. Sweden, PV.652, p. 8; Japan, PV.661, p. 7–8, Sweden, PV.764, p. 16.
106. Argentina, PV.578, p. 8; Sweden; PV.676, p. 9; Sweden, CCD/461, p. 12.
107. Japan, CCD/PV.739, pp. 8–13; Japan, CCD/529.
108. Sweden, CD/PV.29, p. 36.
109. Yugoslavia, PV.714, p. 34; Canada, CCD/PV.740, p. 9; Iran, CCD/745, p. 9.
110. United Kingdom draft convention CCD/512, Article II; United Kingdom, PV.720, p. 7.
111. Sweden, PV.569, pp. 24–25; United Kingdom, draft convention, CCD/512, Article II.
112. Netherlands, PV.560, p. 7; Non-aligned working paper CCD/400, Article III, A.a.
113. German Democratic Republic, CCD/PV.747, p. 17.
114. Sweden, CCD/322, p. 2; Sweden, PV.569, p. 25.
115. United States of America, PV.702, p. 16; United States of America, PV.711, p. 9; United States of America, PV.740, p. 27; United Kingdom, CCD/PV.801, p. 19.
116. United Kingdom, CD/PV.2, p. 61; Federal Republic of Germany, CD/PV.5, p. 34; Netherlands, CD/PV.6, p. 16; Switzerland, CD/PV.31, p. 52; Federal Republic of Germany, CD/PV.29, p. 19; Sweden, CD/PV.29, p. 32; United Kingdom, CD/15; Egypt, CD/PV.31, p. 16; United Kingdom, CD/PV. 29, p. 21; Italy, CD/PV.29, p. 26.
117. United States of America, PV.702, p. 13; United Kingdom, CCD/PV.761, p. 18; United Kingdom, CCD/541.
118. Poland, PV.722, p. 9; Canada, CCD/PV.740, p. 12.
119. Italy, CD/PV.29, p. 25.
120. Australia, CD/PV.31, p. 10.
121. Sweden, PV.569, pp. 24–25; Yugoslavia, PV.569, p. 14; Yugoslavia, CCD/377, p. 1; USSR, PV.583, p. 12.
122. Socialist draft convention CCD/361, Articles IV, V; USSR, PV.567, p. 19; Pakistan, PV.571, p. 19; USSR, PV.583, p. 12; Mongolia, PV.572, p. 9; Socialist Working Paper CCD/403, p. 1; Czechoslovakia, PV.621, p. 9; USSR, PV.714, pp. 30–31.
123. USSR, PV.567, p. 19; Socialist Working Paper CCD/403, p. 1; United States of America, PV.702, p. 13; USSR, PV.704, p. 16.
124. USSR, PV.759, p. 11; USSR, CCD/358; USSR, CCD/740, p. 17.
125. Czechoslovakia, PV.621, p. 9.
126. Japan, PV.631, p. 10; Japan, draft convention CCD/420, Article V.
127. USSR, PV.567, p. 19; Egypt, PV.572, p. 26; Argentina, PV.578, p. 10; Finland, CCD/412, p. 2.
128. USSR, PV.583, p. 12; Mongolia, PV.616, p. 10; Czechoslovakia, PV.621, p. 10.
129. USSR, CCD/522, p. 9; USSR, CCD/PV.728, p. 20.
130. United States of America, PV.702, p. 8.
131. United States of America, CCD/360, p. 6; United States of America, PV.551, p. 21.
132. United States of America, PV.613, p. 13; United States of America, PV.618, p. 8; Netherlands, PV.624, p. 13.

133. Japan, PV.588, p. 11; United States of America, PV.702, p.6; Venezuela, CD/PV.29, p. 9.
134. Federal Republic of Germany, CCD/PV.765, p. 8.
135. United States of America, PV.618, p. 6.
136. Argentina, PV.576, p. 12; Sweden, CCD/375, p. 2; Sweden, PV.590, p. 10; Japan, PV.631, p. 12.
137. United Kingdom, CCD/752, p. 9.
138. Sweden, CCD/395, p. 4; Sweden, PV.590, pp. 10–11; United States of America, PV.618, p. 6; Sweden, PV.622, pp. 6–7.
139. Federal Republic of Germany, CCD/PV.771, p. 23.
140. Federal Republic of Germany, CD/PV.29, p. 17.
141. USSR, PV.593, p. 9.
142. Netherlands, PV.624, p. 13; Sweden, CCD/485, pp. 5–6; United States of America, PV.702, p. 14; Sweden, PV.704, p. 7.
143. Netherlands, CCD/PV.741, p. 19.
144. Sweden, CCD/395, pp. 4–5; Sweden, PV.590. p. 11.
145. USSR, PV.567, p. 19; Socialist Working Paper CCD/403, pp. 1–3.
146. United States of America, CCD/360, p. 10; Sweden, PV.610, p. 13; United Kingdom, draft convention CCD/512, Article VIII; United Kingdom, PV.720, p. 11; USSR, CCD/PV.788, p. 7; United States of America, CCD/PV.802, p. 20.
147. Italy, PV.552, p. 25; Sweden, PV.569, p. 27; Yugoslavia, PV.569, p. 15; Yugoslavia, CCD/377, p. 4; Argentina, PV.578, p. 8.
148. United States of America, CCD/365, p. 10; United States of America, PV.702, p. 16; United Kingdom, draft convention CCD/512, Article III.2.
149. Yugoslavia, CCD/377, p. 3; Sweden, CCD/395, p. 2; Japan, draft convention CCD/420, Article VI-2(a); Egypt, CD/PV.31, p. 15.
150. Sweden, PV.569, p. 27; United States of America, PV.702, p. 16.
151. Non-aligned working paper CCD/400, paragraph 14; Japan, draft convention CCD/420, Article VI; United Kingdom, draft convention CCD/512, Articles VIII, IX, X.
152. Yugoslavia, CCD/377, p. 3; Non-aligned working paper CCD/400, paragraph 14; Japan, draft convention CCD/420, Article VI-2(c), United Kingdom, draft convention CCD/512, Article VIII.
153. Finland, CCD/412, p. 3; Finland, CD/PV.31, p. 50; Finland, CD/14.
154. United States of America, CCD/360, pp. 10–12; Netherlands, PV.560, p. 11; Sweden, PV.589, p. 27; Egypt, PV.572, p. 26; Egypt, PV.603, p. 14; Non-aligned working paper CCD/400, paragraph 14; Netherlands, CCD/410, p. 2; Japan, draft convention CCD/420, Article VI; United States of America, PV.702, p. 16; United Kingdom, draft convention CCD/512, Article VIII; Romania, CCD/PV.743, p. 10.
155. Netherlands, PV.560, p. 10; Sweden, PV.569, p. 27.
156. Sweden, PV.549, pp. 12–13; United States of America, CCD/360, pp. 11–12; United States of America, PV.702, p. 16.
157. Sweden, PV.549, p. 13; Yugoslavia, PV.712, p. 36.
158. United States of America, CCD/360, p. 11; Poland, PV. 611, p. 11; United States of America, PV.613, p. 18; Sweden, PV. 622, p. 10; Japan, PV.631, p. 11.
159. Poland, PV.551, p. 28; USSR, PV.593, p. 9; Non-aligned working paper CCD/400, paragraph 15; United States of America, PV.613, p. 16; Czechoslovakia, PV.621, p. 8; United States of America, PV. 702, p. 12.

160. Sweden, CCD/485, p. 1.
161. Japan, CCD/PV.739, p. 11; Japan, CCD/529, p. 4.
162. United States of America, CCD/360, p. 7.
163. Sweden, CCD/395, p. 3; Sweden, PV.590, p. 11; Sweden, CCD/PV.785, p. 8; Sweden, CCD/569; Italy, CD/5, p. 2; USSR, CCD/538, p. 2.
164. Sweden, PV.463, p. 7; Sweden, PV.569, p. 20.
165. Socialist working paper CCD/403, p. 2.
166. Canada, CCD/334, pp. 1–5; United Kingdom, CCD/502.
167. United States of America, CCD/360, p. 9.
168. United Kingdom, CCD/371; USSR, CCD/538, p. 4.
169. Canada, CCD/334; United Kingdom, CCD/371; United Kingdom, CCD/502.
170. United Kingdom, CCD/308, p. 4.
171. Federal Republic of Germany, CD/PV.29, pp. 17–18; United Kingdom CD/PV.29, p. 22.
172. United States of America, CCD/360, p. 5.
173. USSR, CCD/PV.788, p. 7; United States of America, CCD/PV.802, p. 20.
174. Egypt, PV.555, p. 12; Netherlands, PV.560, p. 8.
175. United States of America, PV.618, p. 10.
176. United States of America, CCD/283; United States of America, CCD/293; United States of America, CCD/360.
177. Sweden, CCD/395; Sweden, PV.590, p. 11; Socialist working paper CCD/403, p. 2.
178. United States of America, PV.740, p. 26.
179. Japan, CCD/301, p. 1; United States of America, CCD/311; United States of America, CCD/368; Italy, CCD/335, p. 2; Japan, CCD/344, p. 3; Socialist working paper CCD/403, p. 2; United States of America, CCD/437, p. 6; United States of America, PV/702, p. 13; USSR, PV.704, p. 17; USSR, PV.714, p. 30; Australia, CD/PV.2, p. 38; USSR, CCD/538, p. 2.
180. United States of America, CCD/437, p. 3; United States of America, PV.702, p. 13.
181. Egypt, CD/PV.31, p. 16.
182. Sweden, PV.549, p. 13; United States of America, CCD/369; Sweden, CCD/384; Japan, CCD/466.
183. Australia, CD/PV.2, p. 38.
184. United States of America, CCD/293; United States of America, PV.702, p. 12.
185. United States of America, CCD/332, pp. 5–6.
186. United States of America, CCD/332, pp. 6–7; Finland, CCD/501; United Kingdom, CCD/502, p. 1; Finland, CCD/577.
187. Netherlands, CCD/PV.748, p. 22; Netherlands, CCD/533; USSR, CCD/538, p. 2.
188. Finland, CCD/412, p. 2; Finland, CCD/453, pp. 2–3.
189. United States of America, PV.702, p. 15; Sweden, CCD/485, p. 6.
190. United States of America, CCD/332; United States of America, CCD/360, p. 8; United States of America, PV.702, pp. 4, 14; United States of America, CCD/498.
191. Poland, CCD/PV.764, p. 12.
192. Federal Republic of Germany, CD/PV.29, p. 18.
193. United States of America, PV.702, pp. 12–13; Poland, PV.722, p. 9.
194. Federal Republic of Germany, CCD/458; Federal Republic of Germany, PV.674, p. 6.

195. Sweden, PV.676, p. 7.
196. Poland, CCD/PV.735, p. 12.
197. United States of America, CCD/366, p. 2.
198. Netherlands, PV.552, p. 17; Egypt, PV.555, p. 12; Netherlands, PV.560, p. 8; United States of America, CCD/366, p. 5; Netherlands, CCD/PV.741, p. 19.
199. USSR, CCD/538, p. 3.
200. Netherlands, CCD/PV.741, p. 16.
201. United States of America, PV.654, pp. 11–12; Sweden, CCD/485, p. 1; United States of America, PV.702, p. 8; USSR, CCD/538, p. 3.
202. Yugoslavia, PV.714, p. 34; United Kingdom, draft convention CCD/512, Article II; United Kingdom, PV.720, pp. 10–11.
203. Netherlands, CCD/PV.758, p. 24.
204. USSR, CCD/PV.788, p. 7; United States of America, CCD/PV.802, p. 20.
205. United States of America, CCD/336, p. 2.
206. Sweden, CCD/485, p. 7.
207. United States of America, CCD/366, pp. 1–3.
208. Sweden, CCD/324; United States of America, CCD/360, p. 6; United States of America, CCD/366; United States of America, CCD/367; Finland, CCD/381; Socialist working paper CCD/403; Canada, CCD/434; United States of America, CCD/436; Sweden, CCD/485; United States of America, CCD/497; United States of America, CCD/498; German Democratic Republic, CCD/506.
209. USSR, CCD/522, p. 9; USSR, CCD/PV.728, p. 20.
210. United States of America, CCD/367.
211. United States of America, PV.654, pp. 11–12; Sweden, CCD/485, p. 1; Poland, CCD/735, p. 12.
212. United States of America, CCD/367, p. 2; United States of America, CCD/436; Canada, CCD/434.
213. Sweden, CD/PV.29, p. 34.
214. USSR, CCD/PV.759, p. 12; USSR, CCD/539.
215. United States of America, PV.654, p. 12; Sweden, CCD/485, p. 7.
216. USSR, CCD/538, p. 3.
217. United States of America, CCD/436; United States of America, PV. 654, p. 12.
218. Japan, draft convention CCD/420, Art. II-3; Japan, PV.631, p. 8.
219. United States of America, PV.702, pp. 12–13; Poland, PV.722, p. 9.
220. Non-aligned working paper CCD/400, paragraph III-c; Sweden, PV.622, p. 9.
221. Sweden, CCD/485.
222. United States of America, CCD/436; Hungary, PV.721, p. 11.
223. USSR, PV.647, p. 19; Sweden, PV.652, p. 10; USSR, PV.652, p. 19; Sweden, CCD/485; Sweden, PV.704, p. 6.
224. United States of America, PV.711, p. 9.
225. Netherlands, CCD/PV.758, p. 22.
226. Sweden, PV.499, pp. 8, 10; Netherlands, PV.560, p. 8; Yugoslavia, PV.377, p. 2; United States of America, CCD/PV.740, p. 24.
227. Netherlands, PV.560, p. 8; Sweden, CCD/PV.764, p. 16; Netherlands, CCD/PV.758, p. 25.
228. United Kingdom, CCD/308, p. 4.
229. Yugoslavia, CCD/PV.742, p. 9.
230. Japan, draft convention CCD/420, Article I-a; United Kingdom, draft convention CCD/512, Article I-a.

231. Yugoslavia, CCD/503, p. 8; Yugoslavia, PV.714, pp. 35–36; United Kingdom, draft convention, CCD/512, Article XII(I); United Kingdom, CCD/PV.761, p. 18; United Kingdom, CCD/541.
232. Egypt, PV.744, p. 11.
233. Sweden, PV.569, p. 25.
234. United States of America, CCD/360, p. 12; Socialist draft convention CCD/361, Articles VI, VI-1, VII; Non-aligned working paper CCD/400, paragraphs 16–17; United Kingdom, draft convention CCD/512, Article X-2.
235. Sweden, PV.549, p. 11; Netherlands, PV.560, p. 11; Sweden, PV.569, p. 25; Pakistan, PV.571, p. 21.
236. Socialist draft convention CCD/361, Article V; Netherlands, PV.560, p. 10; Non-aligned working paper CCD/400, paragraph 13; Japan, draft convention CCD/420, Article VII.
237. United States of America, CCD/360, p. 10; United States of America, PV.702, p. 16.
238. United Kingdom, draft convention CCD/512, Articles VIII, X.
239. Japan, draft convention CCD/420, Article VII.
240. Sweden, CCD/395, p. 3; Sweden, PV.590, p. 11.
241. Japan, draft convention CCD/420, Article IX–1; Japan, PV.631, p. 11.
242. Sweden, PV.499, p. 12; Japan, draft convention CCD/420, Article IX.
243. Sweden, CCD/395, p. 5; Sweden, PV.590, p. 11; Japan, PV.631, p. 11.
244. Non-aligned working paper CCD/400, paragraph 17; Japan, draft convention CCD/420, Article VIII-3; United Kingdom, draft convention CCD/512, Article X–1.
245. Japan, draft convention CCD/420, Article VI.
246. Finland, CCD/381; United States of America, PV.702, p. 16; United Kingdom, draft convention CCD/512, Article X–1.
247. Egypt, PV.572, p. 26.
248. Socialist draft convention CCD/361, Article XI; Japan, draft convention CCD/420, Article XVII; United Kingdom, draft convention CCD/512, Article XIV.
249. United States of America, CCD/360, p. 12; Socialist draft convention CCD/361, Article XI; Japan, draft convention CCD/420, Article XVII; United Kingdom, draft convention CCD/512, Article XIV.
250. United States of America, CCD/360, p. 12; Socialist draft convention CCD/361, Article XI; Japan, draft convention CCD/420, Article XVII–1; United Kingdom, draft convention CCD/512, Article XIV.
251. United States of America, CCD/360, p. 14; Socialist draft convention CCD/361, Article X; Japan, draft convention CCD/420, Article XVI; United Kingdom, draft convention CCD/512, Article XIII.
252. United States of America, CCD/360, p. 13; Socialist draft convention CCD/361, Article IX–1; Non-aligned working paper CCD/400, paragraph 1; Japan, draft convention CCD/420, Article XIV–1; United Kingdom, draft convention CCD/512, Article XII–1.
253. Non-aligned working paper CCD/400, paragraph 5; Venezuela, CD/PV.29, p. 10.
254. Socialist draft convention CCD/361, Article XII–1; Hungary, PV.554, p. 15; Japan, draft convention CCD/420, Article XVIII–1; United Kingdom, draft convention CCD/512, Article XV–1.

255. United States of America, CCD/360, p. 14.
256. Socialist draft convention CCD/361, Article XII–2; Japan, draft convention CCD/420, Article XVIII–2; United Kingdom, draft convention CCD/512, Article XV–2.
257. United States of America, CCD/360, p. 13; Socialist draft convention CCD/361, Article XIII; Japan, draft convention CCD/420, Article XIX; United Kingdom, draft convention CCD/512, Article XVI.
258. Socialist draft convention CCD/361, Article XIII-2; Japan, draft convention CCD/420, Article XIX–2.
259. United Kingdom, PV.720, p. 12.
260. Sweden, PV.569, p. 26; USSR, PV.567, p. 18; Non-aligned working paper CCD/400, paragraph 10; Poland, PV.611, p. 9.
261. Japan, draft convention CCD/420, Article IV; United Kingdom, CCD/PV.741, p. 31.
262. United Kingdom, PV.720, p. 5.
263. Japan, draft convention CCD/420, Article IV; Venezuela, CD/PV. 29, p. 10.

Source: Committee on Disarmament document CD/26, 1 July 1979.

Appendix 3

Parties to the Convention on the prohibition of the development, production and stockpiling of bacteriological (biological) and toxin weapons and on their destruction, as of 1 September 1979

Signed at London, Moscow and Washington on 10 April 1972.
Entered into force on 26 March 1975.
Depositaries: UK, US and Soviet governments.
The total number of parties, as of 1 September 1979, is 84.

Afghanistan	26 Mar 1975	Jamaica	13 Aug 1975
Australia	5 Oct 1977	Jordan	30 May 1975
Austria	10 Aug 1973[a]	Kenya	7 Jan 1976
Barbados	16 Feb 1973	Kuwait	18 Jul 1972[d]
Belgium	15 Mar 1979	Lao People's Democratic Republic	20 Mar 1973
Benin	25 Apr 1975	Lebanon	26 Mar 1975
Bhutan	8 Jun 1978	Lesotho	6 Sep 1977
Bolivia	30 Oct 1975	Luxembourg	23 Mar 1976
Brazil	27 Feb 1973	Malta	7 Apr 1975
Bulgaria	2 Aug 1972	Mauritius	7 Aug 1972
Byelorussia	26 Mar 1975	Mexico	8 Apr 1974[e]
Canada	18 Sep 1972	Mongolia	5 Sep 1972
Cape Verde	20 Oct 1977	New Zealand	13 Dec 1972
Congo	23 Oct 1978	Nicaragua	7 Aug 1975
Costa Rica	17 Dec 1973	Niger	23 Jun 1972
Cuba	21 Apr 1976	Nigeria	3 Jul 1973
Cyprus	6 Nov 1973	Norway	1 Aug 1973
Czechoslovakia	30 Apr 1973	Pakistan	25 Sep 1974
Democratic Yemen*	1 Jun 1979	Panama	20 Mar 1974
Denmark	1 Mar 1973	Paraguay	9 Jun 1976
Dominican Republic	23 Feb 1973	Philippines	21 May 1973
Ecuador	12 Mar 1975	Poland	25 Jan 1973
Ethiopia	26 May 1975	Portugal	15 May 1975
Fiji	4 Sep 1973	Qatar	17 Apr 1975
Finland	4 Feb 1974	Romania	27 Jul 1979
German Democratic Republic	28 Nov 1972	Rwanda	20 May 1975
Ghana	6 Jun 1975	San Marino	11 Mar 1975
Greece	10 Dec 1975	Saudi Arabia	24 May 1972
Guatemala	19 Sep 1973	Senegal	26 Mar 1975
Guinea-Bissau	20 Aug 1976	Sierra Leone	29 Jun 1976
Hungary	27 Dec 1972	Singapore	2 Dec 1975
Iceland	15 Feb 1973	South Africa	3 Nov 1975
India	15 Jul 1974[b]	Spain	20 Jun 1979
Iran	22 Aug 1973	Sweden	5 Feb 1976
Ireland	27 Oct 1972[c]	Switzerland	4 May 1976[f]
Italy	30 May 1975	Taiwan	9 Feb 1973[g]

* This refers to the People's Democratic Republic of Yemen (Southern Yemen).

Thailand	28 May 1975	Union of Soviet Socialist Republics	26 Mar 1975
Togo	10 Nov 1976	United Kingdom	26 Mar 1975[h]
Tonga	28 Sep 1976	United States	26 Mar 1975
Tunisia	18 May 1973	Venezuela	18 Oct 1978
Turkey	25 Oct 1974	Yugoslavia	25 Oct 1973
Ukraine	26 Mar 1975	Zaïre	16 Sep 1975

The following states signed the Convention, but have not ratified it: Argentina, Botswana, Burma, Burundi, Central African Republic, Chile, Colombia, Democratic Kampuchea, Egypt, El Salvador, Gabon, Gambia, Federal Republic of Germany, Guyana, Haiti, Honduras, Indonesia, Iraq, Ivory Coast, Japan, South Korea, Liberia, Madagascar, Malawi, Malaysia, Mali, Morocco, Nepal, Netherlands, Peru, Somalia, Sri Lanka, Syria, United Arab Emirates, United Republic of Tanzania, Yemen (Northern Yemen).

Notes:

[a] Considering the obligations resulting from its status as a permanently neutral state, Austria declares a reservation to the effect that its co-operation within the framework of this Convention cannot exceed the limits determined by the status of permament neutrality and membership with the United Nations.

[b] In a statement made on the occasion of the signature of the Convention, India reiterated its understanding that the objective of the Convention is to eliminate biological and toxin weapons, thereby excluding completely the possibility of their use, and that the exemption in regard to biological agents or toxins, which would be permitted for prophylactic, protective or other peaceful purposes, would not in any way create a loophole in regard to the production or retention of biological and toxin weapons. Also any assistance which might be furnished under the terms of the Convention would be of a medical or humanitarian nature and in conformity with the Charter of the United Nations. The statement was repeated at the time of the deposit of the instrument of ratification.

[c] Ireland considers that the Convention could be undermined if reservations made by the parties to the 1925 Geneva Protocol were allowed to stand, as the prohibition of possession is incompatible with the right to retaliate, and that there should be an absolute and universal prohibition of the use of the weapons in question. Ireland notified the depositary government for the Geneva Protocol of the withdrawal of its reservations to the Protocol, made at the time of accession in 1930. The withdrawal applies to chemical as well as to bacteriological (biological) and toxin agents of warfare.

[d] In the understanding of Kuwait, its ratification of the Convention does not in any way imply its recognition of Israel, nor does it oblige it to apply the provisions of the Convention in respect of the said country.

[e] Mexico considers that the Convention is only a first step towards an agreement prohibiting also the development, production and stockpiling of all chemical weapons, and notes the fact that the Convention contains an express commitment to continue negotiations in good faith with the aim of arriving at such an agreement.

[f] The ratification by Switzerland contains the following reservations:

1. Owing to the fact that the Convention also applies to weapons, equipment or means of delivery designed to use biological agents or toxins, the delimitation of its scope of application can cause difficulties since there are scarcely any weapons, equipment or means of delivery peculiar to such use; therefore, Switzerland reserves the right to decide for itself what auxiliary means fall within that definition.

2. By reason of the obligations resulting from its status as a perpetually neutral state, Switzerland is bound to make the general reservation that its collaboration within the framework of this Convention cannot go beyond the terms prescribed by that status. This reservation refers especially to Article VII of the Convention as well as to any similar clause that could replace or supplement that provision of the Convention (or any other arrangement).

In a note of 18 August 1976, addressed to the Swiss Ambassador, the US Secretary of State stated the following view of the US government with regard to the first reservation: The prohibition would apply only to (*a*) weapons, equipment and means of delivery, the design of which indicated that they could have no other use than that specified, and (*b*) weapons, equipment and means of delivery, the design of which indicated that they were specifically intended to be capable of the use specified. The government of the United States shares the view of the government of Switzerland that there are few weapons, equipment or means of delivery peculiar to the uses referred to. It does not, however, believe that it would be appropriate, on this ground alone, for states to reserve unilaterally the right to decide which weapons, equipment or means of delivery fell within the definition. Therefore, while acknowledging the entry into force of the Convention between itself and the government of Switzerland, the United States government enters its objection to this reservation.

[g] The USSR stated that it considered the deposit of the instrument of ratification by Taiwan as an illegal act because the government of the Chinese People's Republic is the sole representative of China.

[h] The United Kingdom recalled its view that if a régime is not recognized as the government of a state, neither signature nor the deposit of any instrument by it, nor notification of any of those acts will bring about recognition of that régime by any other state. It declared that the provisions of the Convention shall not apply in regard to Southern Rhodesia unless and until the British government informs the other depositary governments that it is in a position to ensure that the obligations imposed by the Convention in respect of that territory can be fully implemented. In a note addressed to the British Embassy in Moscow, the Soviet government expressed the view that the United Kingdom carries the entire responsibility for Southern Rhodesia until the people of that territory acquire genuine independence, and that this fully applies to the BW Convention.

Index